The Structures and Properties of Solids
a series of student texts

General Editor:
Professor Bryan R. Coles

The Structures and Properties of Solids 7

Electron Microscopy in the Study of Materials

P. J. Grundy and G. A. Jones
Lecturers
Department of Pure and Applied Physics
Salford University

Edward Arnold

© P. J. Grundy and G. A. Jones 1976

First published 1976 by Edward Arnold (Publishers) Limited
25 Hill Street, London W1X 8LL

Boards Edition   ISBN 0 7131 2521 7
Paper Edition   ISBN 0 7131 2522 5

All rights reserved. No part of this publication may be
reproduced, stored in a retrieval system, or transmitted in
any form or by any means, electronic, mechanical, photocopying,
recording or otherwise, without the prior permission of
Edward Arnold (Publishers) Limited

Printed Photolitho in Great Britain by
J. W. Arrowsmith Ltd., Bristol.

General Editor's Preface

In most branches of physical science the logical structure into which the subject is broken down for pedagogic purposes is dictated by the different types of *theoretical* concepts most usefully applied to the different parts. This is the case for most of the titles in this series; for the *Crystal Structure of Solids* we start from symmetry elements and their combination, for the *Dynamics of Atoms in Crystals* we are concerned with lattice vibrations, for the *Electronic Structures of Solids* the discussion revolves around the allowed energy levels of electrons.

However there have been some *experimental* developments which have had so great an influence on whole fields of science that a correct picture of the resultant activity and the growth structure of the field requires the identification of the technique and its applications as a topic in its own right, and one deserving of its own books. Electron Microscopy has had such a role in Solid State Physics, and has changed some of its working concepts from theoretical constructs to experimental observables. In fact it has grown in so many directions that it is difficult for the general reader to obtain a broad overview of its range and power. On both these grounds the present book fills a natural position in this series; and should prove useful to a range of readers much wider than the students of formal courses.

As with all the titles in this series care has been taken to make this book reasonably self-contained. Thus, appendices are supplied on various aspects of crystallography which are more fully dealt with in *The Crystal Structure of Solids* by Brown and Forsyth. Similarly the contribution of electron microscopy to studies of defects in solids is illustrated by accounts of a number of such applications which are discussed in a more general context in *Defects in Crystalline Solids* by Henderson.

Imperial College
London
1975

B. R. C.

Preface

This book is intended as an introductory account of the applications of electron microscopy to the study of materials, exclusively solids, which fall within the orbit of the physical sciences. (By implication therefore the widespread use of electron microscope techniques in the biological field is largely neglected.) Primarily it is meant to demonstrate the types of information which may result from electron microscopy and to give a physical basis of the techniques used to obtain it. This being said the book has been written to meet what the authors consider a need in two areas. First as an up-to-date text on the subject for the university undergraduate and polytechnic student of solid state physics, metallurgy and materials science. Naturally the phrase 'up-to-date' must be treated with some caution in an activity which is under continual development. Second it is intended as a guide to those entering the subject for the first time at a practical level, either in industry or postgraduate research. The choice of material for both these purposes has been influenced by the authors' experience gained in lecturing at the undergraduate level and during short advanced 'extramural' courses.

Following a short introduction which lays the historical and scientific context of electron microscopy, the interaction of electrons with matter is considered in Chapter 2. In keeping with the aims of the book a fairly detailed account is presented since electron interactions are crucial to contrast theory. By comparison, the design and construction of electron microscopes is sketched only briefly in Chapter 3 and information about the practical manipulation of microscopes, the design of specimen stages or the art of specimen preparation must be sought elsewhere. The kernel of the book is contained within the next two chapters which deal at length with many of the applications of transmission and scanning electron microscopy. Some of these are well established e.g. the determination of Burgers vectors and the exploitation of topographic contrast, whereas others are either novel or less well known. In the latter category might be classed weak beam methods, Lorentz microscopy and cathodoluminescence techniques. The final chapter looks at more recent trends such as scanning transmission, high voltage and analytical electron microscopy. It can be appreciated therefore that the field of application of electron microscopy is very wide and that an understanding of the principles behind the various techniques calls for a basic knowledge of many branches of physics. In most cases this prior knowledge will be assumed and in any event can be found in other books in this series.

Any book treating the subject of electron microscopy must contain many micrographs and the authors are indebted to those who have sent, helped procure or actually obtained specific material. In particular may be mentioned Dr P. Chippindale and Dr L. J. Rabbitt of Salford University for assistance in supplying many of the micrographs used in Chapter 5. Unfortunately any reproduction process must detract in some measure from the original quality of the plates. Finally we would like to thank Dr L. J. Rabbitt and Dr J. R. Banbury for reading the manuscript sections on conductive mode contrast and scanning transmission microscopy respectively.

SI units are making perceptible headway amongst those who practise electron microscopy and they are therefore employed almost exclusively here. One notable exception is that of the torr (1 mm Hg) which is maintained as the unit of pressure in preference to its SI equivalent. A final point regarding notation. With the proliferation of microscopes available today a system of abbreviations has evolved. While appreciating the merits of such a system, the authors are aware that inconsistencies can result from its use.

The authors would like to thank the following journals and institutions for permission to use published figures; IEEE, IITRI, The Institute of Physics, Journal of Materials Science, Micron, Philosophical Magazine, Physica Status Solidi, Scientific American, The Royal Microscopical Society and The Royal Society.

Sale   PJG
1975   GAJ

Contents

1	INTRODUCTION	
1.1	Resolving power and the light microscope	1
1.2	Electron waves	3
1.3	The development of electron microscopes	4

2	INTERACTION OF ELECTRONS WITH SOLIDS	
2.1	Introduction	8
2.2	Transmission of electrons	10
2.2.1	Incoherent elastic scattering and contrast	11
2.2.2	Coherent elastic scattering and diffraction	17
2.3	Inelastic scattering and energy losses	26
2.4	Electron scattering from bulk material and associated effects	28
2.4.1	Emissive and reflective effects	30
2.4.2	Recombination and cathodoluminescence	35
2.4.3	X-ray and Auger electron emission	36
2.4.4	Charge flow in the specimen	38

3	**THE ELECTRON MICROSCOPE**	
3.1	Introduction	39
3.2	The electron gun	39
3.3	Magnetic lenses and their aberrations	42
3.4	A brief physical description of the CTEM	45
3.5	A brief physical description of the SEM	51
3.6	Considerations of resolution	56
3.6.1	The CTEM	56
3.6.2	The SEM	58
3.7	The merits and disadvantages of electron microscopes	61
4	**CONVENTIONAL TRANSMISSION ELECTRON MICROSCOPY**	
4.1	Introduction	62
4.2	Surface information and external morphology	62
4.3	Electron diffraction and simplified theories of amplitude or deficiency contrast in crystals	68
4.4	Contrast from an imperfect crystal	78
4.4.1	Lattice defects	79
4.4.2	Precipitates and second phases	86
4.5	Specialized techniques in transmission electron microscopy	90
5	**THE SCANNING ELECTRON MICROSCOPE AND ITS APPLICATIONS**	
5.1	Introduction	98
5.2	The reflective and emissive modes	99
5.2.1	Topographical and atomic number contrast	99
5.2.2	Electric and magnetic field contrast	103
5.2.3	Voltage (potential) contrast	105
5.2.4	Electron channelling patterns (ECP's)	107
5.3	Absorbed currents	113

5.4	X-ray microanalysis and Auger spectroscopy	113
5.5	Cathodoluminescence	115
5.6	Induced conductivity	118
5.7	Conclusion: Techniques providing information in specific areas	124

6 RECENT DEVELOPMENTS IN ELECTRON MICROSCOPY

6.1	Introduction	127
6.2	Conventional transmission electron microscopy at high voltages	128
6.2.1	Electron scattering at high energies	128
6.2.2	Instrumental points, aberrations and resolution	131
6.2.3	Areas of application	133
6.3	Analytical transmission electron microscopy	137
6.4	The scanning transmission electron microscope (STEM)	140
6.4.1	High resolution scanning transmission microscopy	144
6.5	Energy analysis and energy analysing microscopes	147
6.6	Conclusion and future trends	152

APPENDIX 154

A.1	The space lattice and its notation	154
A.2	Useful geometrical relationships in a space lattice	156
A.3	Interplanar spacings and angles	156
A.4	Details of the reciprocal lattice	156
A.5	Structure factor	157
A.6	Indexing a polycrystalline ring pattern	158
A.7	Plotting a single crystal pattern	159
A.8	Kikuchi lines and their use in the determination of orientation	161

REFERENCES	165
BIBLIOGRAPHY	167
INDEX	171

List of Main Symbols Used

A	atomic weight
α, β	semi-angular aperture, divergence angle
B	brightness, magnetic induction
C_s, C_c, C	aberration coefficient, contrast
γ	phase shift
d	interplanar spacing, diameter of discs of confusion
e	electronic charge
E	electron energy (kinetic)
F	magnetic flux, force
$f, f(s)$	atomic scattering factor for electrons
η	electron yield or coefficient
θ_B, θ_{hkl}	Bragg angle
i	angle of incidence, current density
$\mathscr{I}$	electric current
I	intensity
K, χ, k	wave vector and wave number
L	camera length, diffusion length
λ	wavelength
Λ	coherence length
M	magnification
m	electronic mass
μ	absorption coefficient
n, n_0	refractive index
N	density of atoms per unit volume, number of raster lines
N_0	Avogadro's number
p	momentum
P	path difference
ρ	density
r	radii of discs of confusion
σ	scattering cross section
s	scattering parameter, deviation from Bragg condition
τ	lifetime

t	thickness, time
T	temperature
V	potential difference, voltage
v	velocity
ϕ	phase difference
X	electric field
ψ	scattering angle or complement of scattering angle
Ψ	wave function
Z	atomic number
ω	frequency
ξ	extinction distance

1

Introduction

1.1 Resolving power and the light microscope

Any device which permits the discernment of detail finer than that which can be seen with the naked eye is of great scientific value. Thus the discovery of the optical microscope in the seventeenth century stimulated a new era of investigation into many aspects of physical and biological science. In succeeding centuries the experimental side of microscopy was brought to a high level of attainment. Complementary developments in the theoretical aspects of microscopy, although arrived at more slowly, were put on a sound footing by the work of E. Abbé. He it was who first explained on the basis of diffraction effects that there is a limit to the smallness of objects which might be reproduced by a lens. This limit is known variously as the resolution, limit of resolution or resolving power of the microscope. Essentially the resolution of the microscope is the smallest separation of two points in the object which may be distinctly reproduced in the image.

Consider the simple optical system shown in Fig. 1.1. Because of diffraction each point in the object is spread out over a small disc — called the Airy disc — in the image. A criterion of resolving power which takes account of this spreading can be expressed quantitatively as $k\lambda/n_0 \sin \alpha$ where λ is the wavelength of the illumination, n_0 is the refractive index in object space, α is the semi-angle subtended by the object at the lens (or lens stop) and k is a constant. The quantity $n_0 \sin \alpha$ is usually called the numerical aperture of the lens. The value of k depends upon the coherence of the illumination but the figure normally adopted in light (and electron) microscopy is 0·61. We may therefore write,

$$\text{Resolving power} = 0\cdot 61\, \lambda/n_0 \sin \alpha \qquad (1.1)$$

It is clear from Equ. 1.1 that the best resolution will be attained using a combination of the shortest wavelength with the largest numerical aperture. The effective wavelength of white light is 550 nm while the highest numerical aperture is about 1·6. Substituting these figures into Equ. 1.1 yields a value for the resolving power of the order of 0·2 μm (200 nm), a performance which had been achieved experimentally in the nineteenth century.

2 INTRODUCTION

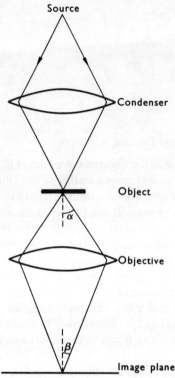

Figure 1.1 A simple optical microscope consisting of condenser and objective lenses.

Beyond this limit there seems no way in which the resolving power of a microscope can be improved. For although electro-magnetic radiations with wavelengths much shorter than those of light do exist, e.g. X-rays, there is no method whereby they can be focussed. Discoveries in the 1920's offered a way out of this apparent impasse however and led eventually to the construction of an electron microscope.

Resolving power is not the only factor which must be considered when the performance of an optical microscope is assessed. It is important that the image should contain sufficient contrast to be discerned against the background. The contrast in the image is usually related to changes in amplitude produced in the specimen on the incident light. Such a specimen is known as an 'amplitude' object as distinct from a 'phase' object which produces local variations in phase. In either case the interaction of the incident light with the object is fundamental. In an electron microscope the interaction of electrons with matter is equally important and this aspect is treated at length in Chapter 2.

1.2 Electron waves

In 1924 L. de Broglie postulated that material particles, e.g. protons, electrons etc., have an associated wave nature, the wavelength of the particle (or a beam of identical particles) being

$$\lambda = h/p \qquad (1.2)$$

where h is Planck's constant and p is the momentum of the particle. For the specific case of the electron, momentum is usually acquired by falling through a potential difference V (in volts), termed the accelerating voltage. Hence, since energy is conserved,

$$eV = \tfrac{1}{2}mv^2 = p^2/2m \qquad (1.3)$$

where e, m, and v are the electronic charge, mass and velocity respectively. The zero of potential is taken to coincide with zero kinetic energy i.e. $V = 0$ for $v = 0$. Equ. 1.3 suggests a convenient method of expressing the energy of an electron or electron beam in terms of electron volts, with

1 electron volt (eV) = $1 \cdot 602 \times 10^{-19}$ J,

being the energy acquired by an electron passing through a potential difference of 1 volt. Equations 1.2 and 1.3 may be solved for λ to give

$$\lambda = h/(2meV)^{1/2} \qquad (1.4)$$

Since V is often very large, the electrons can reach velocities comparable with the speed of light c, and the relativistic increase in mass of the electron should be taken into account. This is most conveniently done by replacing V in Equ. 1.4 by the 'relativistic accelerating voltage' V_r, given by $V_r = V[1 + eV/2m_0c^2]$ where m_0 is the electron rest mass. This correction becomes important for $V \geqslant 10^5$ volts. Table 1.1 shows how λ in picometres varies with V.

Table 1.1 The dependence of electron wavelength on accelerating voltage

kV	λ (pm)
20	8·588
50	5·355
100	3·702
200	2·508
500	1·421
1000	0·872

4   INTRODUCTION

With values of electron wavelength some 10^5 smaller than those in the visible spectrum it might be expected from Equ. 1.1 that the resolution of an electron microscope, presupposing the existence of such a device, might be a similar order of magnitude superior to the light microscope. The question of resolution will be dealt with in Chapter 3: suffice it to say here that the resolving power of the electron microscope does exceed that of the optical microscope but not by a factor of 10^5.

As a consequence of the uncertainty principle the position of each electron in the beam is not known exactly; only a region within which the electron is confined can be specified. The spatial extent of this region (the wave packet) is measured in terms of the coherence length. This is an important quantity because on it depends the possibility of the beam's exhibiting interference phenomena. The coherence length decreases with (a) increasing electron source size and (b) increasing spread of electron wavelengths contained within the beam. If a more coherent beam is required the coherence length should be increased. A typical value for an electron microscope is 20 nm which is very much greater than the electron wavelength. As a rule, if an electron beam of coherence length Λ falls on object points having a spacing less than Λ, these points will be coherently illuminated and coherent interference effects can occur. A fuller discussion of this topic will be found in Heidenreich (1964) and its practical aspects are referred to in Chapter 4.

1.3   The development of electron microscopes

The de Broglie hypothesis, with its subsequent vindication, was one of two decisive advances which led to the development of electron microscopy. The second step was taken by H. Busch in 1926 who demonstrated that axially symmetrical magnetic fields will focus electrons. (Focussing was later shown also for electrostatic fields.) A field distribution which possesses such a focussing action is known as a 'lens' — either magnetic or electrostatic. A brief description of 'magnetic lenses', the ones more commonly used in present-day electron microscopes will be found in Chapter 3.

This discovery led to rapid progress in the field of electron optics and the first electron microscope was constructed by M. Knoll and E. Ruska in 1931. Their instrument was an analogue of the optical microscope (Fig. 1.1) with a resolving power of several tens of nanometres. Since electrons are transmitted through the object this type of instrument has been widely known as the transmission electron microscope, TEM, or just simply as the electron microscope. At the present time to avoid confusion with more recent developments there is a trend to adopt the notation of conventional transmission electron microscope or CTEM, a usage which will be found here.

THE DEVELOPMENT OF ELECTRON MICROSCOPES 5

Following the pioneer work of Knoll and Ruska, various groups in Europe, America and Japan developed the CTEM as a commercial device. A gradual improvement in performance has been achieved although the basic design remains the same and manufacturers now claim resolutions of the order of 0·2 nm (2 Å).

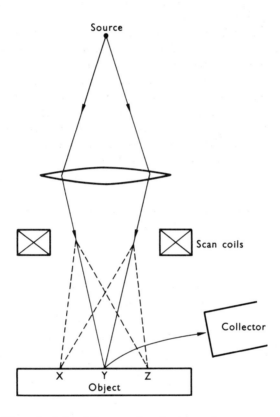

Figure 1.2 The principle of the scanning electron microscope.

A different type of electron optical instrument known as a scanning electron microscope, or SEM, was built by M. von Ardenne in 1938 following a suggestion of Knoll. The principle of the SEM is shown in Fig. 1.2. A source of electrons is focussed by the lens as a spot which is caused to move across the object (X → Y → Z). As the spot strikes each point on the object a response is produced; these responses are collected sequentially and displayed to give an 'image'. Despite its unfamiliar appearance an optical analogy may be drawn between the SEM and a projection light microscope used to observe metallurgical specimens in reflection. One important feature of the scanning microscope is that the image formed is such as if the object were viewed from the electron source, an obvious aid to interpretation. As will be seen later the resolution of the instrument is determined partially by factors not usually associated with the optical microscope and lies at present in the range 0·5–20 nm under ideal conditions.

The history of the SEM has been somewhat erratic. In spite of its early invention, technical difficulties and a lack of the realization of its capabilities hampered its development. Indeed commercial scanning microscopes only became available about 1965, mainly as the result of the pioneer work of C. W. Oatley and his collaborators. Since that time, because of increasing pressure from industrial and research interests alike, manufacture has commenced in many parts of Europe and in Japan. A recent estimate has suggested an investment of about £40 million in the device. Accordingly the whole of Chapter 5 is devoted to the practice and applications of scanning microscopy. The later developments and future trends in electron microscopy will be discussed in Chapter 6.

Finally it seems appropriate, before entering into details of design and techniques, to consider the position of electron microscopy in the context of modern science, especially as there is a tendency to regard any form of microscopy as merely 'the taking of pictures'. Judged by its widespread use in all branches of science and more appropriately by the importance of the information which it has provided (especially in biological science), the electron microscope might be claimed as the most significant scientific instrument invented in the twentieth century. The transmission electron microscope has made its most valued contribution in revealing the crystallographic defects (dislocations etc.) of solids.* As a result the properties and behaviour of these defects have been examined in great detail and the fruits of observation used widely in many aspects of materials science. Compared with the long history of the CTEM the scanning microscope is a newcomer which has yet to be fully proved, its major use so far having been that of a 'super optical' microscope. However this situation will undoubtedly change as its potential value in many other fields becomes appreciated.

This book deals primarily with the application of electron microscope techniques to the general study of solid matter in the physical sciences. The range of material that is susceptible to these probing techniques is very wide,

---

*This role is clearly brought out in the illustrations to *Defects in Crystalline Solids* by B. Henderson, a companion text in this series.

encompassing metals, semiconductors, minerals, fibres and amorphous structures. As will become evident the forte of the electron microscope is its ability to elucidate structural and crystallographic detail. However under favourable circumstances and for suitable specimens information relevant to electrical and magnetic properties can be obtained.

2

Interaction of Electrons with Solids

2.1 Introduction

If a beam of electrons is directed on to a solid target the electrons are either undeviated or they are scattered and are then absorbed, reflected or transmitted. Only if the target is sufficiently thin is a significant fraction of the beam transmitted. Fig. 2.1 is a schematic diagram of the principal interactions that can occur. The events of interest to conventional transmission electron microscopy, scanning electron microscopy and X-ray microanalysis are indicated. It is usual in scanning electron microscopy to employ incident electron beams in the range 1–50 keV and, in certain modes of operation, to study electrons scattered or 'reflected' in a backwards direction and 'emitted' secondary electrons. In conventional transmission electron microscopy, information is obtained from electrons which traverse the specimen and are either undeviated or are scattered forwards into the aperture of an image forming lens. Here the beam energies are usually between 40 and 200 keV in conventional microscopes and between 200 keV and 3 MeV in the recently developed high voltage machines (see Chapter 6).

The developing technique of scanning transmission electron microscopy, which seeks to marry the advantages of the SEM and the TEM, depends on fundamental interactions and principles explained in this chapter but further consideration of this technique is deferred until Chapter 6 as, at the present time, it represents a new technology.

Electron scattering may be either elastic or inelastic. In the first process the electrons do not lose any appreciable energy; only their direction is changed. A small loss occurs because of the change in momentum on scattering from the 'whole' atom. However, because of the disparity in mass of the scattered electron and the atom the loss is too small ($\Delta E/E \sim 10^{-9}$ at aperture angles used in the CTEM) to affect the coherency of the beam. In the inelastic case energy may be transferred to internal degrees of freedom in the atom or specimen in several ways. This transfer may cause excitation or ionization of the bound electrons, excitation of free electrons or lattice vibrations and possibly heating or radiation damage of the specimen. Measurement of these energy losses can

INTRODUCTION 9

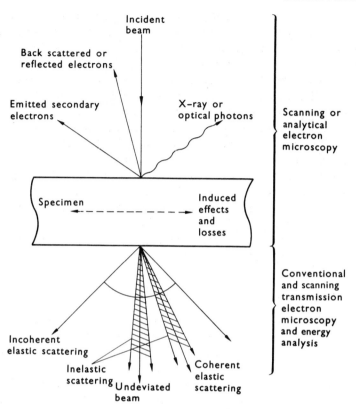

Figure 2.1 A schematic representation of the various interactions of an electron beam with a solid target. The interactions pertinent to the various modes of electron microscopy are indicated.

give information of a chemical nature. Instruments such as the electron probe X-ray microanalyser analyse X-ray emissions and can identify the elements constituting the specimen. This analytical technique will be considered in Chapters 5 and 6 in the context of the use of X-ray spectrometers fitted to the SEM. Energy analysis of transmitted inelastic electrons will be discussed in Chapter 6.

Apart from the distinction between elastic and inelastic scattering it is useful to classify the degrees of scattering into several principal divisions. First, there is single scattering which, as its name implies, indicates that an electron suffers only one interaction; this type of scattering is relevant to the CTEM where very thin films or crystals are studied. Next there are plural or multiple scattering events where an electron may undergo, say, ten or more than twenty successive scattering events respectively. Finally there is 'diffusion-like' scattering where electron motion becomes random. In a thick, bulk target, such as often observed in the SEM, all the above rather arbitrarily defined types of scattering may occur.

The function of this Chapter therefore is to consider the various electron scattering mechanisms that can occur in some detail and to point out the information that can be obtained from them. The sections are arranged in such a way as to separate the areas of interest in conventional transmission and scanning electron microscopy although, fundamentally, the scattering mechanisms are often the same for both applications. For reasons which are to some extent historical, transmission effects are considered first. This is probably the preferable order because there is no doubt that the fundamental factors and theories governing the formation of contrast in CTEM images are fairly well understood. The explanation of contrast in SEM images is, as yet, a more qualitative and empirical one. However, with the rapid advances and popularity of this expanding technique this situation is quickly changing.

2.2 Transmission of electrons

In the CTEM the image is a true microscopic image in the conventional sense in that it is formed from the object by the use of a lens. In light microscopy differences in absorption in the specimen are normally responsible for image contrast whereas in the case of electrons contrast is obtained by utilizing differences in scattering power. In the usual operational mode of the CTEM these differences in scattering power are used to produce 'deficiency' or amplitude contrast. However, as in light microscopy of transparent specimens, the phase differences imparted to the scattered electrons can also be detected and a phase contrast image constructed. Examples of objects which are well suited to the formation of 'phase' images in the CTEM are those which scatter weakly, i.e. at low angles, such as magnetic domains. Images containing phase information can also be constructed from crystalline and non crystalline specimens using special imaging modes. The mechanisms which produce contrast in phase objects are fairly complex and are briefly discussed later in this Chapter. Particular examples are dealt with in Chapter 4. Most weight is given here to a consideration of deficiency contrast as this underlies the most widely used techniques in practice.

Electron scattering is a strong function of angle and so the aperture of the imaging lens is an important parameter. Fig. 2.2 shows the important elements of a simple image forming system. A parallel beam of electrons is incident on a 'thin' specimen situated in the object plane. The transmitted electrons pass

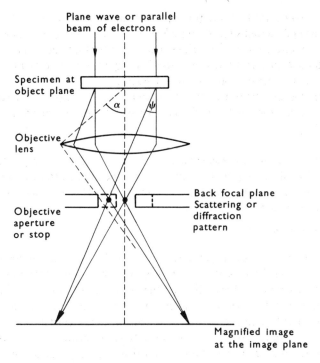

Figure 2.2 A simple optical or electron optical imaging system containing the necessary elements of such a scheme.

through the objective lens (a light optical analogue is shown) and an image is formed in the conjugate, Gaussian image plane of the lens. An aperture can be inserted in the back focal plane of the lens and it can be used as a stop to limit the semi-angular aperture, α, of the lens (the significance of this aperture for resolution is considered in Chapter 3). As can be seen from the schematic ray paths, a point in the focal plane contains electrons scattered at the same angle ψ by different points in the specimen and a particular point in the image contains electrons from a corresponding point in the specimen. The intensity distribution below the specimen and in the back focal plane and its dependence on ψ is of great importance in what follows.

2.2.1 Incoherent elastic scattering and contrast

Consider a monoenergetic electron beam incident normally on a thin solid specimen. As yet no assumption as to the thickness of the specimen is made except that a significant fraction of the incident beam is transmitted. The object is an ideal one and contains scattering atoms of one kind arranged completely at random throughout the specimen. The scattering of the electrons is completely elastic and incoherent, i.e. there is no loss of energy by the beam and the scattering events are random and in no way related.

In a manner analogous to the consideration of photons passing through an absorptive medium, the total transmitted intensity, I, can be written as

$$I = I_0 e^{-\sigma_s t} \tag{2.1}$$

where t is the thickness of the specimen and I_0 the incident intensity. The scattering action of the object is described by σ_s, the total scattering cross section for the material. σ_s is related to the scattering cross section or the effective area of an atom in scattering, σ_a, by

$$\sigma_s = N\sigma_a = N_0 \rho \sigma_a / A$$

Here N is the number of atoms per unit volume, N_0 is Avogadro's number, ρ the density of the material and A the atomic weight of the atoms comprising the specimen. It is found experimentally that the ratio σ_a/A changes only slightly with electron energy for a given specimen, so that the product ρt, often termed the 'mass thickness', largely determines the transmitted intensity.

To study the angular distribution of the scattered electrons it is necessary to consider the differential cross section $D(\psi)$ which is the dependence of σ_a on angle ψ through

$$D(\psi) = \frac{d\sigma_a}{d\omega} = \frac{1}{2\pi \sin \psi} \left(\frac{d\sigma_a}{d\psi} \right) \tag{2.2}$$

and is the fraction scattered into unit solid angle at some angle ψ, (see Fig. 2.3). $D(\psi)$ may be written as $(f(s))^2$ where $f(s)$ is the atomic scattering factor or amplitude. The atomic cross section for scattering between limits ψ and $\psi + d\psi$ is given by

$$\sigma_a = 2\pi \int_{\psi}^{\psi + d\psi} (f(s))^2 \sin \psi \, d\psi \tag{2.3}$$

The intensity of an electron image formed by accepting a particular angular fraction of scattered electrons into the aperture of the image forming lens can, in principle, be calculated from equations of the form of Equ. 2.3. This is done by evaluating the integral and then summing the single atom cross sections for the solid. Approximations to this procedure are considered later and in Chapter 4.

Electrons are scattered by the atomic potential $V(r)$, determined by the nuclear and electron cloud Coulomb potentials. If the kinetic energy E of the

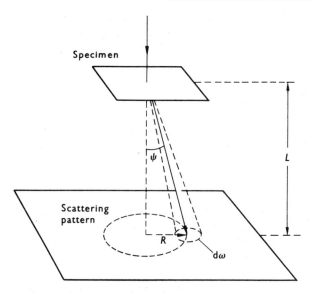

Figure 2.3 Showing the scattering of transmitted electrons by a thin specimen into a solid angle dω at an angle ψ.

incident electrons is large such that $E \gg V(r)$ use can be made of an approximation due to Born where the scattering is 'weak' and the incident electron wave is substantially unmodified. The approximation allows a wave mechanical calculation for $f(s)$, the details of which may be found in the literature, e.g. Pinsker (1953). The result obtained for an atom of atomic number Z is

$$f(s) = \frac{8\pi^2 me^2}{h^2} \left(\frac{Z - f_x(s)}{s^2} \right) . \qquad (2.4)$$

$f_x(s)$ is the corresponding scattering factor for X-rays, which is dependent only on the charge of the electron cloud and s is the so-called 'scattering parameter' = $(4\pi/\lambda) \sin(\psi/2)$. The first term in Equ. 2.4 is the Rutherford scattering formula. Substitution of typical values in Equ. 2.4 shows that for small values of ψ the scattered intensities for electrons and X-rays are related as $D(\psi) \sim 10^4 D_x(\psi)$ and so electrons are scattered far more effectively than X-rays. For real isolated atoms $D(\psi)$ and $f(s)$ decrease monotonically with angle, unlike ideal point scatterers where the scattered intensity is constant with angle. This difference arises from the interference of wavelets scattered at various distances from the centre of the atom with a corresponding increase in phase difference with angle.

14 INTERACTION OF ELECTRONS WITH SOLIDS

For a completely disordered solid a summation of such profiles should give a similar form for the scattered intensity as a function of angle. It can be shown (Heidenreich 1964) that the intensity $I(\psi)$ at a radial point in the scattering pattern in the back focal plane is given by

$$I(\psi) = \left(\frac{I_0}{L^2}\right) D(\psi) \qquad (2.5)$$

per scattering atom in the approximation where ψ is small. Hence for a specimen of thickness t with a density N of atoms of one species

$$I(\psi) = \left(\frac{I_0}{L^2}\right) NtD(\psi). \qquad (2.6)$$

L is an effective distance from the specimen to the back focal plane, Fig. 2.3, and it is clear from Equ. 2.6 that the pattern is a profile of $D(\psi)$. This function falls off monotonically with ψ (as $\sin^{-4} \psi/2$) and therefore as ψ varies between 0 and $\pi/2$ the scattered intensity should fall off smoothly. Strictly speaking the scattering pattern or, as considered later, the diffraction pattern is only formed in the back focal plane when a collimated beam is incident on the specimen. For convergent or divergent illumination the poorly defined pattern is formed above or below the focal plane. This point should be noted as perfect collimation is rare and is only approached in certain special conditions. It is observed in practice that many materials which are considered to be totally disordered, i.e. amorphous in the true sense, produce scattering patterns which exhibit diffuse, subsidiary maxima at $\psi > 0$ as in Fig. 2.4. This arises from the fact that total disorder is rare and in a liquid or amorphous solid the number and distance of nearest neighbours to a given atom are not random but cluster around preferred values often related to those in the crystalline solid.

This primitive state of order requires that the electrons be treated as waves. The subsidiary maxima are a result of interference between scattered waves and in this situation the scattered intensity and the differential cross section are given by

$$I(\psi) \propto D(\psi) = N(f(s))^2 \left(1 + \frac{1}{N} \sum_{n \neq n'} \sum \frac{\sin sr_{nn'}}{sr_{nn'}}\right) \qquad (2.7)$$

The first term in this equation is the monotonically decreasing radial intensity already discussed. The second term, the Debye function, is the 'structure sensitive intensity' and is an angular function of s (with $r_{nn'}$ the distance between atoms n and n') which superimposes maxima and minima on the pattern. As the degree of order increases and 'micro-crystallites' are present the subsidiary maxima become sharper. When the specimen can be considered to be crystalline,

TRANSMISSION OF ELECTRONS 15

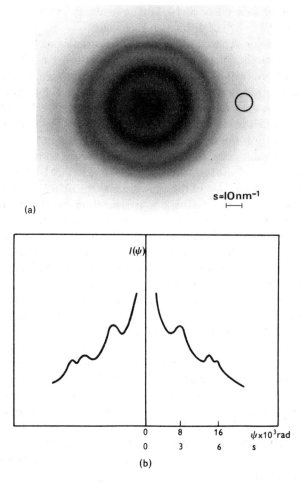

Figure 2.4 Electron scattering by a thin, amorphous cobalt film; (a) the scattering pattern and a representation of the objective aperture and (b) the scattered intensity $I(\psi)$ as a function of ψ or s.

i.e. the regions of order in it are considerably greater in size than the resolving power of the imaging system, the scattering is usually treated in terms of diffraction theory, as explained in the following sub-section. However, it should be realised that this approach is essentially an extension of the ideas considered here as the Debye function is valid for the crystalline case; thus for a unique interatomic distance r, $\sin sr/sr$ makes a maximum contribution to $I(\psi)$ when tan $sr = sr = 2\cdot 46\pi$ or $\lambda = 1\cdot 6r \sin(\psi/2)$. The similarity of this last relation to the Bragg equation (Equ. 2.8) of diffraction is obvious.

Contrast is usually obtained in the image by restricting part of the scattering pattern in the back focal plane by use of an aperture stop (see Fig. 2.2). This practice is discussed further in later chapters. Deficiency contrast is observed by removing the electrons scattered at angles greater than the semi-aperture angle of the system, α. The diameter of the stop or objective aperture defines α. In truly amorphous materials with complete disorder no contrast is observed if $\sigma_s t$ remains constant over the specimen because no one point in the specimen scatters to any greater extent than another. In specimens with a small degree of order the experimental situation is usually the same in fact because the widths of the subsidiary maxima are of the same order as the objective aperture diameter, as shown in Fig. 2.4. This is equivalent to saying that no detail is observed in the image if it is smaller than the resolving power of the system, which is of the order of $0\cdot 2$ nm in a modern CTEM. However, as will be discussed in Chapter 4, by using some of the techniques common to optical microscopy it may be possible to discern detail in 'non-crystalline' specimens in the form of localized interference fringes which have been associated with very small regions of local order. These phase contrast images are obtained in a defocussed mode of operation and make use of interference effects between waves of varying phase as referred to above.

The maximum thickness of a disordered specimen, e.g. a carbon film, a carbon or plastic replica or a non-crystalline metal or glass film, that can be conveniently used in transmission can be defined by the approximate condition $0\cdot 5 < \sigma_s t < 1$. For 50 keV electrons the maximum thickness of carbon ($A = 12$) is ~ 200 nm and platinum ($A = 195$) is ~ 15 nm. These figures correspond to a critical mass thickness of 3×10^{-4} kg m^{-2}. Approximate calculations of this kind are only valid for single scattering, but plural or multiple scattering where electrons can be scattered back into angles $\psi < \alpha$ do not change the essential criteria of contrast.

Contrast is observed experimentally as a fractional change in transmitted intensity with position in the image. The parameters defining contrast are t and σ_s. If I_b is the background intensity given by Equ. 2.1, contrast is defined as $C = (I - I_b)/I_b$. For a change in t due to a change in effective thickness or an inclination or a protuberance in the specimen

$$C = I_0(e^{-\sigma_s(t+\Delta t)} - e^{-\sigma_s t})/I_0 e^{-\sigma_s t} = 1 - e^{-\sigma_s \Delta t} \simeq \sigma_s \Delta t \quad \text{for} \quad \sigma_s \Delta t < 1.$$

The scattering pattern is unmodified in shape by the protuberance, only the overall intensity is increased and the contrast is defined by the additional scattering outside the objective aperture defined by $\sigma_s \Delta t$. This situation is illustrated in Fig. 2.5. For a superimposed or included particle of scattering cross

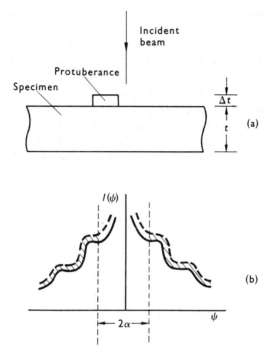

Figure 2.5 Contrast in an amorphous specimen due to a protuberance. The protuberance is illustrated in (a) and the curve in (b) shows the extra scattering from the feature outside the objective angular aperture.

section σ_p, C is obtained as $\sigma_p \Delta t$ and $(\sigma_p - \sigma_s)t$ respectively. The minimum value of C that can be conveniently detected photographically is about 5% so that rough limiting values of Δt or $\Delta \sigma_s$ in the above situations can often be calculated. Smaller values of C can be detected in many situations by dark field measurements. The experimental arrangement for dark field microscopy is considered in Chapter 3.

Finally it should be pointed out that measurement of the scattering pattern enables the atomic arrangement and state of order to be investigated in the case of non-crystalline solids such as glasses, amorphous semiconductors and metals. Reference should be made to James (1950), Guinier (1963) and Giessen and Wagner in Beer (1972).

2.2.2 Coherent elastic scattering and diffraction

As indicated above, specimens that are crystalline in the accepted sense that they contain crystal grains greater in size than about 2–10 nm must be thought

18   INTERACTION OF ELECTRONS WITH SOLIDS

of as producing scattered coherent electron waves. These waves have definite phase relationships with each other and with the incident electron wave and diffraction and subsequent interference effects are obtained over large angles. Crystalline specimens produce regular electron diffraction patterns and image contrast which cannot be explained by the scattering of particulate electrons, by mass thickness concepts or conveniently by Equ. 2.7.

The requirements for electron diffraction can be illustrated by reference to Fig. 2.6. Plane electron waves in the monoenergetic incident beam are thought of as being 'reflected' from planes of atoms in the crystal rather than scattered from individual atoms. This is the Bragg model for diffraction. The path

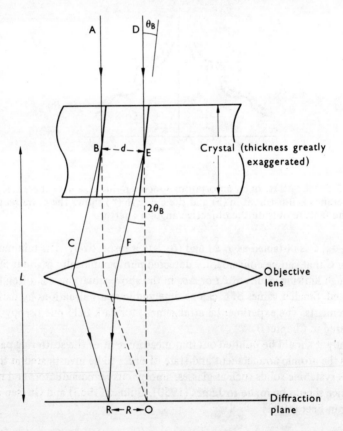

Figure 2.6   Illustrating schematically Bragg reflection from planes in a thin crystal and the formation of an interference maximum at a point in the back focal plane of the objective lens.

difference between the parallel beams or waves ABC and DEF after reflection at B and E is PEQ = $2d \sin \theta$, where d is the separation of the reflecting planes and θ is the angle the waves make with the planes. For BR and ER to interfere constructively at R, i.e. to have their peak amplitudes in phase, the condition

$$n\lambda = 2d \sin \theta \qquad (2.8)$$

or Bragg's law must hold, where λ is the electron wavelength and n is an integer (unity for a first order reflection). The interplanar spacing is characteristic of a particular set of planes in the crystal and is often written as d_{hkl} where $h\ k$ and l are the Miller indices which define the plane. The calculation and properties of Miller indices are considered in the Appendix. The value of θ that satisfies Equ. 2.8 for particular values of d and λ is the Bragg angle which is often written as θ_B or θ_{hkl}. Typical values of d and λ are 0·2 nm and 4 pm (for 100 keV electrons) respectively, which gives θ_B as about $1°$. It is thus clear that the planes responsible for diffraction are almost parallel to the incident beam. This result is important for electron microscopy.

As in the case of the 'scattering' patterns of the previous section, the diffraction pattern is formed near the back focal plane of the imaging lens in a manner analogous to the Fraunhofer diffraction pattern from an optical grating. It can be inferred from Figs. 2.3 and 2.6 that the interference maximum or diffraction 'spot' or 'ring' at R is related to the Bragg angle θ_B by a relation of the form $R = 2\theta_B L$ for small θ. L is an effective distance from the object to the diffraction pattern and is often called the diffraction or camera length. Using Equ. 2.8 it is possible to write

$$R = \frac{L\lambda}{d} \qquad (2.9)$$

for small θ and it is apparent that distances in the diffraction pattern are in inverse proportion to distances in the specimen. This is an important result because it justifies the introduction of the concept of a lattice in inverse relation to a real crystal lattice. This reciprocal lattice is extremely useful in predicting the form of diffraction patterns and in their interpretation. Some properties of the reciprocal lattice are given in the Appendix where the interpretation of diffraction patterns is also outlined.* The important point in principle is that the reciprocal lattice is composed of a system of scattering points each of which corresponds to a reflecting plane in the real lattice and has the same Miller indices as the corresponding plane. Fig. 2.7 shows diffraction patterns from a

---

*See also *The Crystal Structure of Solids* by P. J. Brown and J. B. Forsyth, No. 2 in this series.

20 INTERACTION OF ELECTRONS WITH SOLIDS

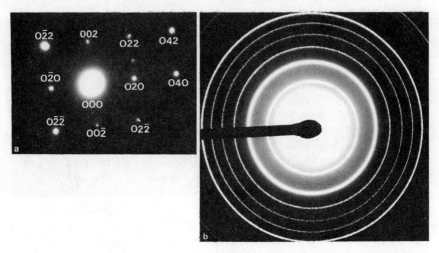

Figure 2.7 Diffraction patterns from (a) a thin single crystal of nickel and (b) a polycrystalline film of thallous chloride.

Figure 2.8 (a) Below. Illustrating schematically the formation of Kikuchi lines in transmission electron diffraction. (b) Opposite. A typical pattern.

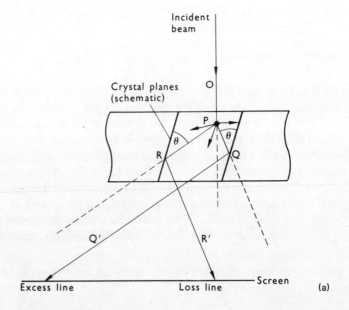

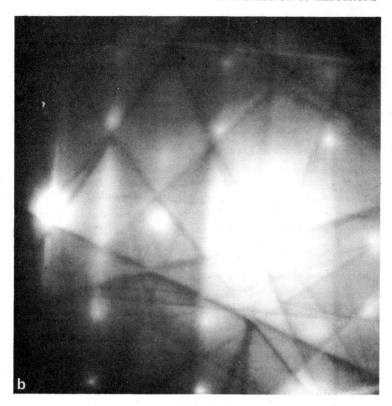

single crystal of nickel and a polycrystalline film of thallium chloride. Each spot in the nickel pattern is labelled with the Miller indices of its plane of origin. For the polycrystalline specimen the spots degenerate into concentric rings (passing through the positions of the spots) because the incident electron beam 'illuminates' many crystals whose diffracting planes are at all azimuthal angles to the beam (this material is often used as a standard for calibration of the camera length).

An effect often observed in electron diffraction patterns of fairly thick and flat crystals is the presence of straight bright or dark lines or bands running across the pattern. These Kikuchi lines originate from inelastically scattered electrons which are subsequently reflected by planes which are at the Bragg angle to those electrons. Fig. 2.8 illustrates this schematically; some electrons in the

22 INTERACTION OF ELECTRONS WITH SOLIDS

incident beam are inelastically scattered at P. Those which reach the reflecting planes at R and Q are Bragg reflected into RR' and QQ'. Because PQ is nearer the forward direction (i.e. more nearly along OP) than PR the intensity of the electrons scattered into QQ' is greater than that of RR'. A reciprocal lattice construction shows that the intersection of QQ', the excess or bright Kikuchi line, on the screen approximates to a straight line. The defect line, a line reduced in intensity from the background intensity of the diffraction pattern, also has a line intersection on the screen. To each reflecting plane or spot in the diffraction pattern there is thus a pair of Kikuchi lines. When the crystal is tilted the spots stay still and change only in intensity but the Kikuchi lines move across the pattern. This is explained by considering Figs. 2.6 and 2.8. As the planes B and E or R and Q tilt about the Bragg angle associated with the fixed direction of the incident beam, the reflection will go in and out of register but the direction of the diffracted beam, if present, will be constant (it will be shown below that a change of intensity can occur over a range $d\theta_B$ of the Bragg angle.) However the inelastic electrons from P in Fig. 2.8 strike R and Q over a range of angles and as R and Q are tilted there are always some electrons incident at θ_B and a Bragg reflection is obtained over a wide range of tilt. As the direction of the electrons selected for reflection changes there is a corresponding change in direction of the reflected beam and a movement of the Kikuchi lines across the pattern. Kikuchi lines can be used as an aid in analysing electron diffraction patterns and image contrast as indicated in the Appendix.

The points X and Y of Fig. 2.9 may be considered as two reciprocal lattice points separated by a vector r. The incident and diffracted waves are described by reciprocal wavelength vectors or wave numbers K and K_d ($|K| = |K_d| = \lambda^{-1}$). The path difference P between the waves scattered at X and Y is XB − AY or

$$P = \frac{r}{|K|} \cdot (K_d - K) \tag{2.10}$$

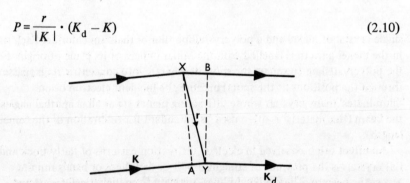

Figure 2.9 The construction for calculating the path difference between the incident wave and that scattered from a reciprocal lattice point at r.

TRANSMISSION OF ELECTRONS 23

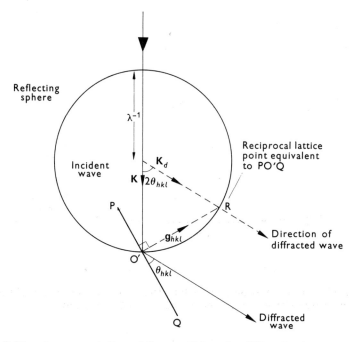

Figure 2.10 A representation of the conditions for diffraction in terms of the reciprocal lattice and the Ewald reflecting sphere.

where $P = n\lambda$ for a diffraction maximum. The conditions for diffraction from a plane can be obtained by relating the reciprocal lattice to the Ewald 'reflecting' sphere, Fig. 2.10. A sphere of radius λ^{-1} is drawn through the origin of the reciprocal lattice (at O' on the plane PQ) so as to intersect a point R which is the reciprocal lattice point corresponding to PQ. The incident beam is a diameter of the sphere and is incident on PQ at O'. The reciprocal lattice vector g_{hkl} defines the particular reflection at an angle θ_{hkl} from PQ. In this situation $\sin \theta_{hkl} = g_{hkl} \lambda/2 = \lambda/2d_{hkl}$, i.e. Bragg's law is satisfied and diffraction occurs. It is clear that $K_d - K$ is identified here with g_{hkl} and defines the normal to the reflecting plane. In conventional transmission electron microscopy the diameter of the reflecting sphere is ~ 500 nm^{-1} (i.e. $\lambda \sim 4$ pm) and a typical reciprocal lattice distance is $1/d_{hkl} \sim 5$ nm^{-1}. Hence the reflecting sphere approximates to a plane surface with respect to the reciprocal lattice and can intersect many reciprocal lattice points as shown in Fig. 2.11. The diffraction pattern of a crystal, which is the projection of the reciprocal lattice in the plane normal to the incident beam, can therefore consist of many spots or reflections.

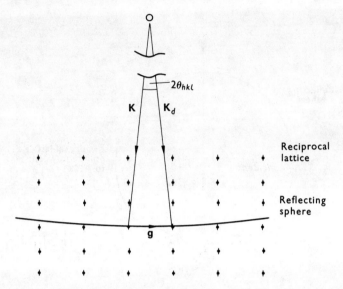

Figure 2.11 An illustration, in two dimensions, of the intersection of the reflecting sphere with many 'extended' reciprocal lattice points.

The fundamental equations describing the conditions for electron diffraction and the geometry of diffraction patterns are those due to von Laue who originally discussed them in the context of X-ray diffraction. Equ. 2.10, written as $r \cdot K/|K| = n\lambda$, is the Laue condition. It is apparent that for a three dimensional lattice of scattering points with lattice directions defined by repeat distances a, b and c that the three Laue conditions for a diffraction maximum are

$$a \cdot K = h, \quad b \cdot K = k \quad \text{and} \quad c \cdot K = l \qquad (2.11)$$

h, k and l are the Miller indices for the particular reflection in the absence of any common factor and all three conditions must be satisfied simultaneously. In practice the Bragg equation is a more convenient expression to use in diffraction work and it can be shown quite rigorously that it is an alternative way of expressing the more esoteric Laue equations (see e.g. Brown and Forsyth (1973)).

As indicated in the previous section, and as will be discussed further in later chapters, the principal method of obtaining contrast in the CTEM is by obstructing part of the electron distribution below the specimen. In crystalline specimens this deficiency contrast is often called 'diffraction' contrast because all the diffraction pattern save one beam is obstructed. A complete understanding of the details of contrast requires a knowledge of the intensities of diffraction from different parts of the specimen as well as the geometry of diffraction. A calculation of the intensities involves the phase difference ϕ between diffracted waves. ϕ is obtained from Equ. 2.10 as

$$\phi = \frac{2\pi}{\lambda} P$$
$$= 2\pi r \cdot (K_d - K)$$
$$= 2\pi(r \cdot g_{hkl}) \tag{2.12}$$

at the Bragg position. This result may be written in terms of real and reciprocal lattice cell dimensions (see the Appendix) as

$$\phi = 2\pi(ua + vb + wc) \cdot (ha^* + kb^* + lc^*)$$
$$= 2\pi(uh + vk + wl) \tag{2.13}$$

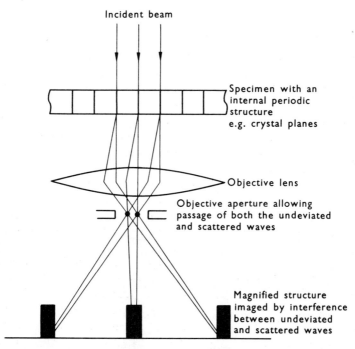

Figure 2.12 An illustration of the scheme used to image lattice planes in crystals and in phase contrast microscopy in general. Only a one-dimensional periodic structure is considered.

The general calculation of diffraction intensities and some specific examples are treated in Chapter 4.

In the diffraction contrast mode any information concerning the arrangement of atoms in the lattice of the crystal is lost. This is because image detail from the periodic part of the object is only obtained, according to the Abbé theory, if the image is reconstructed by interference between the directly transmitted beam and one or more of the diffracted beams. This means that to obtain such image detail several beams must be accepted through the objective aperture to interfere in the image plane. In the case of a crystal this produces a periodic set of fringes that can be associated with the crystal planes producing the diffracted beam(s). This mode is explained schematically in Fig. 2.12 and its applications in the resolution of lattice planes are discussed in Chapter 4. This method of producing contrast is the same as that suggested earlier for disordered materials. It is sufficient to say here that very ideal specimen and instrumental conditions are required for success in this imaging mode whereas the diffraction contrast technique has almost universal application in crystalline materials.

2.3 Inelastic scattering and energy losses

Inelastic scattering involves the loss of energy from the electron beam to the specimen. The types of loss that can occur have already been listed and it is the function of this section to discuss the relevance of these losses to the transmission and scanning modes of electron microscopy.

Losses such as ionization and excitation of atoms can occur in free atoms but solids have additional inelastic cross sections that are 'collective' and are the result of the binding together of atoms in the solid state. The most important of these collective losses is that of plasma excitation of electrons. The incident electrons cause the valence electrons or conduction electrons, in the case of metals, to oscillate with a characteristic plasma frequency ω_p associated with a quantum mechanical 'plasmon' of energy $\Delta E = h\omega_p/2\pi$, where h is Planck's constant. The energy of the incident electron is reduced by ΔE and it is deflected through a small angle. The characteristic energy losses are of order 5–30 eV for most solids and can have sharp values, e.g. 5·5 eV for SiO_2, or broad spectra, e.g. in transition metals. This scattering is not localized and the production of a plasmon is indeterminate so that the scattered electron carries no information on the physical structure of the specimen (unless it is subsequently elastically scattered) and gives a diffuse contribution to any diffraction pattern or image. By calculating the scattering cross section for plasma oscillations it can be shown that electrons, initially accelerated by 100 kV, which lose energy equivalent to 15 eV are scattered through an angle of $\sim 10^{-4}$ rad. This is a hundred times smaller than a typical Bragg angle so these electrons fall well within the aperture of the image forming system in the CTEM. The loss electrons have maxima coincident with Bragg reflections so that diffraction spots are surrounded by an inelastic contribution. The mean free path for the formation of a plasmon is ~ 100 nm and therefore an appreciable fraction of transmitted electrons from a typical specimen of thickness of order 100 nm is inelastic. The inelastic fraction decreases at high beam voltages and small specimen thicknesses and is the factor which essentially limits the thickness of crystalline specimens suitable for the CTEM. Little will be said about another collective loss, that of phonon excitation, as here the energy loss is very small ($\lesssim 1 \text{eV}$) and the scattered electron is virtually elastic. The mean free path for phonon excitation is calculated as being of the order $1-5$ μm so that it is not a very probable event in transmission electron microscopy. What losses do occur contribute a diffuse background to elastic scattering and impair image contrast.

Non-collective events such as excitation or ionization of bound atomic electrons in inner levels have relatively small cross sections in thin sections as compared to plasma excitation, at least in specimens composed of light elements, so that these interactions are not a very serious source of inelastic scattering as far as imaging in the CTEM is concerned. The scattering angles involved are generally greater than those for plasmon scattering and the 'lossy' electrons tend to degrade images at the atomic scale. However there are important implications for bulk specimens in scanning electron microscopy and these are discussed in the next section.

In all the significant loss processes so far considered energy is given to the specimen although energy gain is possible for phonon scattering. One general result of this is that the temperature of the specimen must be affected. Specimen heating is, of course, greater in non-conducting specimens where heat cannot escape to the rest of the apparatus. In metals, significant temperature increases can be avoided by providing a specimen support which conducts the heat away and increases of temperature greater than about $10°C$ can be avoided at normal working beam currents. Beam induced heating can be a serious problem in organic materials where degradation of chemical bonds can occur. Use of low beam currents is a help in avoiding such troubles. Radiation damage in specimens and the displacement of atoms can also occur. This effect is normally reduced at low beam currents. The effects of radiation damage can be serious in materials containing light atoms such as carbon and hydrogen where the energy transferred to an atom can easily be greater than the displacement energy. Displacement effects in metals are only observed at high beam currents and above certain beam energies corresponding to critical energies required for the displacement of atoms in the specimen as discussed in Chapter 6.

28   INTERACTION OF ELECTRONS WITH SOLIDS

It is obvious from the foregoing that energy loss processes can be investigated in transmission (in both the CTEM and STEM) not only to assist in the reduction of their detrimental effect on image quality but also as a source of information on the specimen. Collective and localized losses are typical of a particular material and can be used to identify the chemical composition and to investigate some physical properties of the specimen. Instruments have been developed for this purpose and are known as energy analysing microscopes; they usually take the form of modified electron microscopes or microscopes with additional, 'modular' type equipment. As yet the applications of the techniques of analysing 'lossy' electrons have been limited mainly to specific areas of research and commercially available instruments are not common as yet. This situation must change in the future. Some discussion of electron energy analysis is given in Chapter 6. The important application of X-ray analysis in the SEM from localized losses is introduced in the next section and discussed further in Chapter 5.

2.4   Electron scattering from bulk material and associated effects

The aim here is to describe the scattering mechanisms and associated effects that are pertinent to the explanation of image contrast in scanning electron microscopy. 'Bulk' specimens are usually investigated and hence there is little probability of electron transmission (there is now, however, some interest in transmission in the so-called STEM as discussed in Chapter 6) and attention centres on the production of 'back' scattered and emitted electrons from the near surface regions of the specimen, Fig. 2.1. The incident electrons experience a greater path length in the specimen than in a transmission microscope specimen and the probability of inelastic events such as ionization is increased as well as plural, multiple and diffusion scattering. Consequently the theory of scattering in bulk specimens becomes very complex although many models to explain it have been proposed: for an account see Wells (1974). As far as scanning microscopy is concerned the important factors are how the incident electrons lose energy, their range of penetration into the specimen and their angular deviation.

Fig. 2.13 represents schematically the 'pear shaped' volume into which the incident electrons are dispersed. This spreading of the beam inside the surface has an important bearing on resolution of the SEM (see Chapter 3). It does not follow that the products of scattering collected outside the target arise from a region identical to the interaction volume. For example, secondary electrons created at the bottom of this volume possess insufficient energy to escape to the

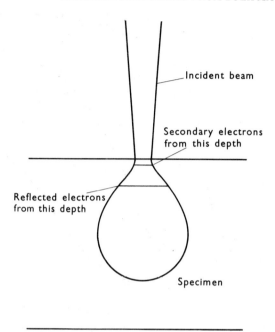

Figure 2.13 A schematic representation of the spreading of an electron beam inside a 'bulk' specimen. The source regions for secondary and backscattered (reflection) electrons are indicated.

surface and so will not be sampled. In general the sampled volume will be less than the interaction volume.

Various definitions of 'range' have been proposed, itself a testimony to the complexity of the scattering problem. One definition, that of extrapolated range, can be obtained from results such as those shown schematically in Fig. 2.14. These represent the fractional transmission of normally incident electrons through films of various thicknesses as a function of primary incident energy. If the linear portion of a curve corresponding to thickness x' is extrapolated to cut the abscissa at E'_0 then x' is called the extrapolated range R_x for incident primary energy E'_0. As might be expected R_x increases with E'_0. A useful quantity is the mass-range, ρR_x (analogous to the mass thickness concept for transmission microscopy) which for a given value of E_0 is approximately independent of atomic number. Some results from Cosslett and Thomas (1964) are presented in Table 2.1, from which it should be possible to obtain some idea of the penetration into any given specimen.

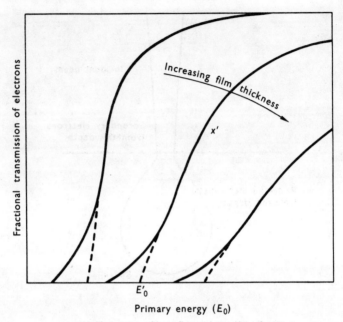

Figure 2.14 The definition of extrapolated range. The graph is schematic only.

Secondary electrons, products of ionization by in-going or out-going 'reflected', primary electrons form a substantial fraction of the signal coming from the specimen, the remainder being reflected electrons many of which have energies close to the incident primary energy.

The following subsections deal in rather more detail with scattering events that may occur within the interaction volume.

Table 2.1 The mass range (ρR_x) in μg cm^{-2} as a function of the incident electron energy for different target elements. The atomic number of each element is given in brackets (after Cosslett and Thomas, 1964)

E_0	2.5 keV	5 keV	10 keV	15 keV
Al (13)	36	90	230	—
Cu (29)	43	102	305	530
Ag (47)	42	96	280	510
Au (79)	40	97	280	510

2.4.1 Emissive and reflective effects

The principal modes of contrast in the SEM depend upon the detection of two categories of electrons. First, the reflected primaries or back scattered electrons with energies 50 eV $\lesssim E \leqslant E_0$ and second, emitted secondaries with energies $\lesssim$ 50 eV but greater than the work function of the specimen surface. Primaries are scattered from atoms within a shallow depth in the specimen ($\sim$ 1 μm) and must be scattered singly or multiply at large angles to re-emerge from the specimen. Secondary electrons are obtained as an ionization product and because of their relatively low energies must be formed at very shallow depths ($\sim$ 0·1 μm) if they are to scatter or diffuse to the surface and escape from the specimen. Fig. 2.15 is a schematic representation of the electron emission coefficient or electron yield, defined as the number of emitted electrons per incident electron, as a function of the reduced accelerating energy E/E_0. The peak near $E/E_0 = 1$ is the most probable reflected primary energy which is changed from E_0 because of extremely small energy losses ($\sim$ 1 in 10^5) in such events as phonon excitation. The preferred scattering mode at large angles is that of elastic scattering.

The angular distribution of back scattered primaries at normal incidence is

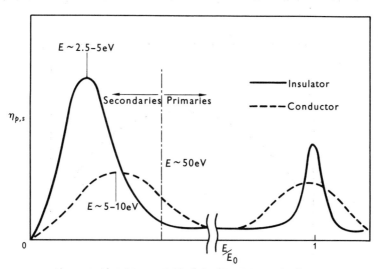

Figure 2.15 Schematic curves for the yield of backscattered and secondary electrons as a function of the reduced energy for conducting and insulating materials. The division of energies for secondary and primary electrons is arbitrary.

32 INTERACTION OF ELECTRONS WITH SOLIDS

given by an empirical relation of the form

$$N(\psi) = \frac{\eta_p}{\pi} \cos \psi \qquad (2.14)$$

where ψ is the complement of the scattering angle, $N(\psi)$ is the number per unit solid angle and η_p is the back scattered coefficient. Clearly the intensity of scattered primaries is observed to increase with the scattering angle. A theoretical justification for many of the experimental observations on primary back scattering is not yet complete because of the complexities introduced by multiple scattering. The more extensive treatments of Thornton (1968) and Sandström *et al* (1974) consider the approaches that can be made in some detail. Experimentally η_p is found to be virtually constant with changing E_0 and to increase with atomic number of the specimen. This result leads to the possibility of compositional differentiation in a specimen, i.e. the detection of different phases. For example, η_p for aluminium is about 0·17 and for uranium it is 0·5.

In practice it is usually the secondary electron fraction that is harnessed in image formation using emission effects. The secondary electron yield η_s is usually greater than η_p and has different orders of value for conductors and insulators (possibly due to differences in electron–electron scattering in the specimen), i.e. for insulators $1 < \eta_s < 20$ and for metals $0·5 < \eta_s < 2$. In general,

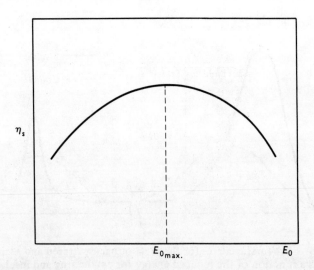

Figure 2.16 The secondary electron yield as a function of incident electron energy (schematic).

η_s is dependent on the reduced energy E/E_0 and E_0 as shown schematically in Figs. 2.15 and 2.16. Fig. 2.16 shows a maximum in η_s for a particular value of E_0. This behaviour is general; the increase in η_s occurs because the production of secondaries at higher beam energies is more probable and the subsequent decrease occurs because the average depth at which they are formed increases and passage to the surface is more difficult. $E_{0\,\text{max}}$ is typically 500 eV for metals and 1–2 keV for oxides and insulators. The effect of the work function on secondary electron energies is illustrated in Fig. 2.15; the smaller values of this parameter in semiconductors and insulators allow lower energy electrons to escape from the surface. In most ordered, i.e. crystalline, materials there is a decrease in secondary emission with temperature increase, probably because of increased scattering but this decrease is only slight.

An important result for scanning electron microscopy is that η_s at a point is found to vary with the angle of incidence i of the electron beam as

$$\eta_s(i) \propto \exp[A(1 - \cos i)] \tag{2.15}$$

A is a constant composed of an absorption term and the average depth from which escaping secondaries reach the surface. As i increases, Fig. 2.17, η_s increases and thus any detected signal is a function of the inclination of a particular object point to the beam. This forms the basis of topography detection as suggested in Fig. 2.18 and discussed in Chapter 5. Equ. 2.15 can be

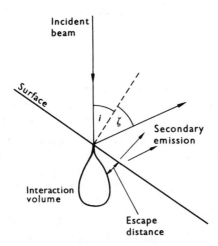

Figure 2.17 An illustration of how the escape distance is reduced for a surface inclined to the incident beam.

34 INTERACTION OF ELECTRONS WITH SOLIDS

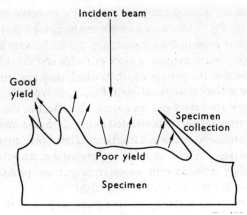

Figure 2.18 Secondary electron emission at exaggerated surface irregularities. The phenomenon of specimen collection is also indicated.

approximately expressed as

$$N(i) = N(0) \sec i \qquad (2.16)$$

for the number of secondaries as a function of i where $N(0)$ is the number per unit time obtained at normal incidence. Topographical contrast obtained in an SEM image is a function of many beam and instrumental parameters, as discussed in Chapters 3 and 5, but the dependence on specimen inclination can be defined as a contrast C given by

$$\frac{d(N(i))}{di} = C = \tan i \, di \qquad (2.17)$$

The minimum detectable difference in surface inclination is then $di = C_{min} \cot i$ where C_{min} is a minimum observable contrast of, say, 5% (see Section 2.2.1). For $i \sim 60°$, $di \sim 1°$ and hence quite small differences in surface roughness are easily detected. On specimens with large scale roughness and re-entrant surfaces contrast can also be obtained by surface 'collection' effects as indicated in Fig. 2.18. However this is often obtained at the expense of detail. If primary electrons are collected differences in surface inclination of about $0\cdot5°$ can be observed since primaries are specularly reflected.

Equations 2.15 and 2.16 deal with the secondary electron yield for a given angle of incidence on to the specimen. Regardless of the orientation of the specimen the secondary electrons which emerge from any point have an angular distribution which closely follows a cosine law (cf. Equ. 2.14 for reflected primaries). For an incident angle i, (see Fig. 2.17), the number $N(i, \zeta)$ of secondary electrons per unit solid angle appearing at angle ζ is given by

$$N(i, \zeta) = \frac{N(0)}{\pi} \cos \zeta \qquad (2.18)$$

Expressed in terms of χ rather than ζ and using Equ. 2.16 this relation becomes (with $\chi = i + \zeta$)

$$N(i, \chi) = \frac{N(0)}{\pi} [\cos \chi + \sin \chi \tan i] \qquad (2.19)$$

Any potential difference between the detector of electrons and the specimen will also produce contrast as the resultant electric field will affect the energies and trajectories of the emitted electrons. This effect can be used to investigate potential distributions in semiconductors as discussed in Chapter 5.

Compositional information of a rudimentary nature can be obtained from both primary and secondary electrons as the respective yields vary from material to material. This compositional information is normally superimposed on topographic detail but can be separated if a differential detection technique is employed.

2.4.2 Recombination and cathodoluminescence

Many of the secondary electrons which are produced by the incident beam do not escape from the target material but diffuse through the specimen and are captured. If the capture process (e.g. electron-hole recombination) is accompanied by the emission of a photon at optical or near optical frequencies cathodoluminescence occurs. Examples of common cathodoluminescent materials are phosphors used in cathode ray tubes and television screens etc., plastics, organic materials and semiconductors.

Radiative recombination, by definition, results in the emission of photons of frequency determined by $\Delta E = h\nu$ where ΔE is the energy released in the process. Cathodoluminescence can be localized or not depending on the ability of the secondaries to diffuse before recombination. Sometimes non-radiative recombination can occur and energy is given to create, say, a phonon in a crystalline solid. For an efficient radiator the average lifetime of 'radiative' secondaries must be much less than the lifetime of non-radiative secondaries. In general the ratio of these two lifetimes can depend on such factors as impurity contents, temperature and the intensity or excitation level of the incident beam. In cathodoluminescent crystals such as Ge, Si, GaAs and ZnS which have well known energy band structures the recombination process is fairly well understood.

As far as contrast is concerned, the local intensity of cathodoluminescence is often associated with surface topography and indeed surface irregularities can affect the light contrast. A complementary study of topography by secondary electron emission is often useful. Contrast is also expected at internal defects, such as dislocations, and at precipitates, as these present regions of extra surface and different recombination and scattering conditions. At positions where the lifetime ratio previously mentioned is different, contrast is also revealed, e.g. at $p-n$ junctions between regions containing different impurities in semiconductors.

2.4.3 X-ray and Auger electron emission

The simple properties of X-rays and details of their production are so widely known and so extensively discussed that they will be given only brief mention. However, as indicated in the introduction to this chapter, and Section 2.3, these properties can be utilized in specimen microanalysis and are therefore of some relevance to electron microscopy. X-ray emission occurs when an inner shell electron is excited by the primary electron beam to leave the atom entirely or go into a higher unoccupied level. The vacancy is subsequently filled by another electron which drops from a higher energy level and emits an X-ray photon of energy $h\nu$ equal to the energy lost ΔE by the electron in falling between the two shells.

Clearly from the microscopic analysis point of view the most vital aspect of X-ray emission is that each element possesses a unique X-ray spectrum, see e.g. Brown and Forsyth (1973). Characteristic X-ray wavelengths and energies are shown in Fig. 2.19 as a function of atomic number. The discrete part of the spectrum, i.e. the presence of sharp $K\alpha$ and $K\beta$ lines corresponds to electronic transitions between L and K shells and M and K shells respectively and L lines result from M to L transitions. The continuous spectrum comprises X-ray photons arising from inelastic collisions between electrons in the beam and those of the target. The presence of the unique X-ray spectrum thus serves two purposes; it allows for an analysis of constituent elements in a target material and also provides a potential source of image contrast. More details of the detection and analysis of X-rays in the SEM together with considerations of resolution will be found in Chapter 5.

A process similar to the production of X-rays and one which has also been harnessed in the SEM is that of Auger electron emission. Again the starting point is ionization. Subsequent rearrangement of the remaining electrons in the atom occurs and a certain amount of energy released. Whereas in the X-ray case the end result is a short wavelength photon, here the energy is used to eject

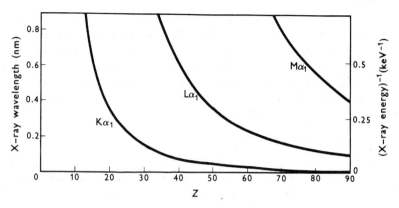

Figure 2.19 Some prominent X-ray wavelengths and energies as a function of atomic number Z.

another, Auger, electron from the atom. Since the atoms of any species have a particular set of energy levels it follows that there will be a spectrum of Auger emission energies characteristic of the particular atom. Measurements of the Auger energies which lie in the range 5–1000 eV affords a method of specimen analysis. Further, if a particular electronic energy is selected information can be obtained about the distribution of the material giving rise to those Auger electrons. Catalogues of Auger spectra are now available.

2.4.4 Charge flow in the specimen

So far in this section attention has been given to back scattered electrons, emitted electrons or photons as carriers of information about a specimen. It will be appreciated that some of the secondary electrons created by the impact of the incident beam remain in the specimen. If an electrical lead is now attached a current will flow to earth in order to neutralize any charging effects. This current which can be made to flow in an external circuit, and can be amplified and used as a signal is given by,

$$\mathscr{I}_a = \mathscr{I}_p - \mathscr{I}_s - \mathscr{I}_r \qquad (2.20)$$

where $\mathscr{I}_p$ is the incident primary current, $\mathscr{I}_r$ the reflected current and $\mathscr{I}_s$ the secondary emitted current. The charge flow is known as the specimen absorbed current or neutrality current. Its magnitude is rather low and of the same order as $\mathscr{I}_p$. For this reason the incident beam current must be fairly high if good contrast is to be obtained. In metal samples absorbed currents can yield useful information complementary to that obtained from the emitted electrons.

However in certain materials, notably semiconductors, a completely different type of charge flow can occur which far outweighs any neutrality effects. When the beam strikes a semiconductor it will produce many electron–hole pairs somewhat like the related phenomenon of photoconductivity. The rate of generation of pairs produced is given by

$$\Delta n = \Delta p = (\mathscr{I}_p/e)(E_0/E_i) \tag{2.21}$$

where n, and p are the number of electrons and holes created respectively per unit time and E_i is the energy required to ionize the host atoms. If an electric field is applied across the specimen the two types of charge separate and move towards opposite electrodes. This movement of charge is known as beam induced conductivity (BIC), and can give rise to pulses of current which may be detected in an external circuit. The current produced is often called the charge collection current. Charge collection micrographs give point to point information about the electrical properties of the sample, as any factors which influence the movement or lifetime of separated electron–hole pairs is likely to produce contrast. The maximum value of the beam induced current is of the order of $10^3 \mathscr{I}_p$, which is clearly in excess of the specimen absorbed current.

In certain circumstances there is no need to apply an external voltage to the sample; this is when the specimen contains 'built in' fields which will separate out the charges. The most obvious example is that of the p–n junction which even in a state of zero bias possesses a depletion region with an associated electric field. The phenomenon of charge collection or induced conductivity has been widely exploited in the investigation of semiconductor materials and devices.

3

The Electron Microscope

3.1 Introduction

Electron microscopes are costly and complex machines whose manufacture draws upon the resources of, amongst others, mechanical engineering, electronics and vacuum technology. It is therefore neither possible nor appropriate in a text of this nature to discuss their design and construction in any detail. More comprehensive sources which deal with specific topics will be mentioned where necessary. Nevertheless some account of the basic electron optical principles and equipment involved must be given in order to appreciate the type of information derived from electron microscopy and how it is obtained. Only principles will be given rather than data about particular commercial models.

Although since 1931 great developments in electron optical devices have taken place it should be apparent from what has already been said that demand has crystallized for two particular instruments, the conventional transmission electron microscope and the scanning electron microscope. It is with these two microscopes that this section is concerned although the scanning transmission electron microscope (STEM) is likely to be a future growth area (Chapter 6). Despite fundamental differences of operation there are features common to both. Each possesses a metal column comprising an electron gun and an assembly of magnetic lenses. The column is evacuated to a pressure of better than 10^{-4} torr in order to decrease the perturbing effects of collisions of the electrons in the beam with air molecules. The low pressure also improves the lifetime of the electron gun filaments. A disadvantage of conventional pumping systems is that hydrocarbon vapour is introduced into the column which when adsorbed on to the specimen decomposes in the electron beam and deposits a contaminant layer on the specimen.

3.2 The electron gun

A typical electron 'triode' gun used for the production of an electron beam is shown in Fig. 3.1. It consists of a tungsten hairpin filament acting as cathode mounted in a (Wehnelt) cylinder which contains a small aperture. The anode is a

THE ELECTRON MICROSCOPE

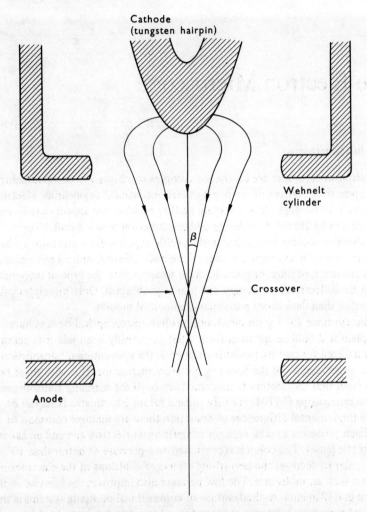

Figure 3.1 A simple triode electron gun showing the formation of the crossover.

flat plate, usually earthed, with an axial aperture through which the electrons pass. A large negative potential (usually 100 kV but greater in particular instruments) is applied between the filament and anode while the cylinder is biased slightly negative with respect to the filament. The electrons are emitted thermionically from the filament with low energies (< 1 eV) before being accelerated and 'focussed' by the electric field of the gun to form a crossover situated somewhere between the cylinder and anode. This crossover, some 50 μm in diameter, and through which all the emitted electrons pass acts as a source for the remaining optical system. The shape of the filament can affect the beam coherence (Section 1.2).

The electron gun plays a very important role in determining the ultimate performance of an instrument (particularly the SEM). The fundamental quality of a microscope gun is the electron brightness which is defined as the current density per unit solid angle. It is given by

$$B = i_c/\pi\beta^2 \tag{3.1}$$

where i_c is the current density at the crossover (Fig. 3.1) and β is the semi-angular aperture and should remain constant for successive images provided the potential is constant. Equ. 3.1 may be written in terms of the absolute current of the electron beam $\mathscr{I}$ and the diameter of the crossover d: for a spot of uniform intensity (only an approximation because the brightness is less at the edges) then

$$\mathscr{I} = \pi^2 d^2 \beta^2 B/4 \tag{3.2}$$

There is an upper limit to B imposed by the maximum current density, i, which may be focussed into a spot. The value of i is given by the Langmuir formula

$$i = i_0 eV\beta^2/kT_0 \tag{3.3}$$

where i_0 is the maximum current density at the filament, V and T_0 the voltage and temperature of the filament respectively, e the electron charge and k is Boltzmann's constant. Substituting Equ. 3.3 into Eq. 3.1 gives the upper limit of B as

$$B = i_0 eV/\pi kT_0 \tag{3.4}$$

Equ. 3.4 shows that the brightness of the gun depends upon filament characteristics. Typical values for tungsten filaments are $T_0 = 2700$ K, $i_0 = 2 \times 10^4$ A m^{-2} and $B = 10^9$ A m^{-2} sr^{-1} for 100 kV. Values of B obtained in practice are normally less than those predicted by Equ. 3.4 which gives the theoretical maximum. To improve brightness the cathode must be modified or redesigned so that a higher value of i_0 can be obtained. Developments along these lines include the use of lanthanum hexaboride filaments ($B = 10^{10}$ A m^{-2} sr^{-1}) and field emission sources. The latter, which are mentioned further in Chapter 6, introduce a practical problem in that they must be surrounded by a region of ultra-high vacuum.

42 THE ELECTRON MICROSCOPE

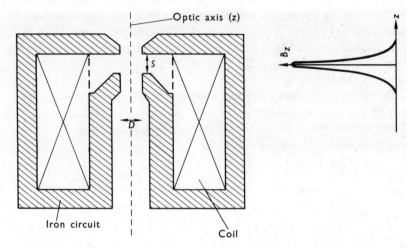

Figure 3.2 A schematic representation of a typical magnetic lens. The component of induction along the optic axis, B_z, is shown as a function of z.

3.3 Magnetic lenses and their aberrations

For several reasons magnetic lenses are preferred to their electrostatic counterparts in most modern microscopes. (An exception is the electron gun which is invariably an electrostatic lens.) They are more reliable, offer no complication of electric breakdown and have better aberration characteristics. A typical example is shown in Fig. 3.2 consisting of an iron circuit energized by a coil, the iron having a gap with parallel pole faces. The component of magnetic induction along the optic axis, B_z, may achieve substantial proportions. The lens properties for a given electron energy can be specified in terms of three parameters: the bore diameter of the pole pieces D, the pole spacing S and the product of the coil windings and excitation current NI. One advantage of the lens is that its focal length may be changed over a certain range by altering the excitation current.

Magnetic lenses may be classified into two groups depending upon their use in the microscope, projector or intermediate lenses and objective lenses. The detailed electron optical distinction between them is beyond the scope of this book. Suffice it to say that for the objective, immersion elements are of interest (the specimen is actually placed inside the lens) whereas asymptotic elements are of interest in the projectors. Fig. 3.3 indicates a ray path through a lens and defines the focal length for projector operation. The plotting of electron trajectories through lens fields is dealt with in the specialist literature. A list of

MAGNETIC LENSES AND THEIR ABERRATIONS 43

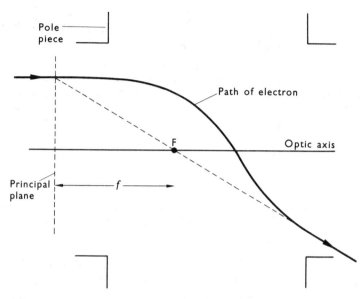

Figure 3.3 An electron trajectory through a magnetic lens. F is the (asymptotic) image focus and f the focal length when the lens is used as a projector.

useful references is given in the Bibliography. If the design of lenses is an esoteric pursuit the study of the aberrations from which they suffer is of more immediate appeal. This is especially true of the CTEM where aberrations have an important bearing on resolution and contrast.

Because of aberrations each point of the object is not imaged as a point but as a disc of confusion. Different aberrations give rise to discs of different sizes each depending on the beam divergence. Although defects occur concurrently it is usual to consider each separately.

(i) *Spherical aberration* arises because the further electrons travel from the optic axis, the more they are bent by the lens and so are focussed at different points along the optic axis: the paraxial rays are focussed at the so-called Gaussian image plane (see Fig. 3.4). As a result each point in the object is imaged as a disc of radius $MC_s\alpha^3$ at the Gaussian image plane where M ($= \alpha/\beta$) is the magnification of the lens. Referred back to object space the radius of the disc is

$$r_s = C_s\alpha^3 \tag{3.5}$$

where C_s is the spherical aberration constant. Actually C_s is not a constant but depends upon the position of the object: it is generally tabulated for the lens

44 THE ELECTRON MICROSCOPE

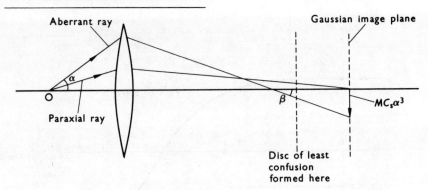

Figure 3.4 The effect of spherical aberration is to image a point O in the object as a disc of radius $MC_s \alpha^3$ at the Gaussian plane.

working in a high magnification mode. For a good objective lens C_s will have a value of about 3 mm but this increases with focal length. Spherical aberration is the most serious aberration encountered in lens design and is unfortunately practically impossible to eradicate. An interesting fact is that the disc of confusion in image space having the smallest size does not occur at the Gaussian plane but is displaced some distance along the optic axis towards the lens. The disc of least confusion has a radius of $\frac{1}{4}MC_s\alpha^3$ in image space.

(ii) The focal length of a magnetic lens depends upon the electron energy and the lens excitation current. Any variation in these parameters will therefore result in a blurring of the image, a defect known as *chromatic aberration* because the electrons are no longer effectively monochromatic. The radius of the disc of confusion due to chromatic aberration is

$$r_c = C_c \alpha \left[\left(\frac{\Delta V}{V} \right)^2 + \left(\frac{2\Delta I}{I} \right)^2 \right]^{1/2} \tag{3.6}$$

where C_c is the chromatic aberration constant, typically 3 mm for an objective lens. ΔI represents fluctuations in lens current while ΔV may arise from instabilities in the accelerating voltage, a spread in electron energies at the filament due to thermal effects or correspond to inelastic scattering at the specimen. The lens and voltage supplies are made as stable as possible to counteract these tendencies.

(iii) Diffraction effects which constitute a fundamental limit to the resolution of light microscopes (Section 1.1) produce a (Airy) disc of confusion whose radius referred to object space is

$$r_d = 0.61\lambda/\alpha \tag{3.7}$$

When Equ. 3.7 is compared with Equ. 1.1 it will be seen that for $n_0 = 1$ and α small, r_d coincides with the resolution criterion adopted in light microscopy.

(iv) Other aberrations such as astigmatism and distortion do exist but are less important in practice because they can be adequately controlled. An aberration peculiar to magnetic lenses is image rotation, which occurs because electrons follow a helical path as they pass through the lens. If diffraction work in the CTEM is contemplated this effect must be allowed for and hence calibrated: experimental details may be found in Hirsch et al. (1965) and Kay (1965).

When all these defects (except those mentioned under (iv)) are taken into account the overall aberrant disc in object space has a size given by

$$r_{aber}^2 = r_s^2 + r_c^2 + r_d^2 \tag{3.8}$$

On the credit side, magnetic lenses have a large depth of field and a large depth of focus, both features arising from the small apertures that define the electron beam. A large depth of field means that a thin object ($\sim 2\mu$m) in a CTEM may be completely in focus for any particular lens setting. The large depth of focus means that the position of the photographic plate or viewing screen is not critical. Quantitatively the two parameters can be written as

$$\left. \begin{array}{l} \text{Depth of field } = 2r/\alpha = 2r'/\beta M^2 \\ \text{Depth of focus} = 2rM^2/\alpha = 2r'/\beta \end{array} \right\} \tag{3.9}$$

where r and r' are the radii of the discs of confusion in object and image space respectively. It will be seen from Equ. 3.9 that the two quantities are related by the factor M^2. In the SEM, the image of the electron source produced by the final lens coincides with the specimen surface and this remains in focus for a range of displacement along the optic axis given by the depth of focus = diameter of final probe/β. In this case a large depth of focus means that rough surfaces can be examined. Texts which deal with the properties of magnetic lenses and their aberrations are listed in the Bibliography.

3.4 A brief physical description of the CTEM

The evacuated microscope column contains an electron gun together with an assembly of lenses. After leaving the gun filament the electrons form a crossover, a demagnified image of which is projected onto the specimen by two condenser lenses. A ray diagram in Fig. 3.5 shows how the first condenser C1 forms a

46   THE ELECTRON MICROSCOPE

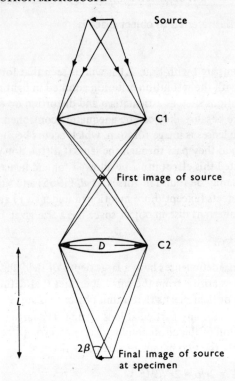

Figure 3.5 A ray diagram of the double condenser system. The beam divergence, β, and the spot size are controlled by varying the strength of C2. The maximum value of 2β is D/L obtained when the beam source is imaged on the specimen. The main advantage of the double condenser is that a small spot size can be produced thus keeping contamination of the sample to a minimum.

demagnified image of size $\sim 1\mu$m which is subsequently projected onto the specimen by C2 with a magnification of about 2x. Thus the final illumination spot on the specimen may be as small as 2μm; this can fill the screen at highest magnification.

The current density at the specimen depends upon filament characteristics and the divergence angle β (Equations 3.1 and 3.4). β itself depends upon the geometrical properties of the condenser system and the excitations of condensers C1 and C2. Some variation in β is allowed by providing an interchangeable series of different size condenser apertures. Typical values for the condenser aperture and the operating divergence angle would be 400 μm and 10^{-3} rad respectively for a 100 kV machine. The first instruments constructed had only a single condenser stage. Although the introduction of a second such lens does not increase the maximum current density available at the specimen it does provide advantages, principally (a) a finer control on the area which may be illuminated so that neighbouring areas do not contaminate and (b) a beam divergence which can be made comparatively low for an equivalent amount of defocussing. A smaller value of β means that the effective electron source size is reduced. This increases the coherence length of the beam which can lead to improved contrast and resolution in images and diffraction patterns.

When the beam strikes the specimen, which must be thin in the CTEM, a number of complex scattering processes can occur; these have already been discussed in Chapter 2. The specimen is usually supported on a fine mesh and placed within some sort of holder. Many types of specimen holder are available, each designed for specific experiments or movements, e.g. goniometer, heating, straining stages etc. A good discussion of the present state of stage design is given by Valdré and Goringe in Valdré (1971).

Many of the scattered electrons which pass through the object now enter the objective lens whose design and aberrations critically affect the performance of the whole microscope. The ray diagram of Fig. 3.6 pertains to a crystalline object and shows coherent Bragg diffracted beams leaving the specimen. An intermediate image I_1 is formed of comparatively low magnification. Thereafter the intermediate lens P1 produces a second intermediate image I_2 which is finally magnified and thrown on to the viewing screen by the projector lens P2. The viewing screen consists of a metal plate with a coating of zinc phosphor which scintillates when struck by the beam thus allowing the image to be observed directly by the eye. Information can also be recorded with a camera placed below the viewing chamber, the screen being lifted to allow electrons to hit the photographic plate or film. The image is focussed on the screen by varying the focal length of the objective lens.

The final magnification is varied by altering the excitations of P1 and P2 and a continuous range of magnification is available from about 1000x to 100 000x and beyond. Calibration can easily be done with a replica grating (Hirsch *et al.*, 1965). Some instruments have a third projector lens which allows more flexibility of the magnification range obtainable (up to 300 000x). Others have inter-changeable projector pole pieces for a similar purpose. The column will also possess some capability for reducing condenser and objective astigmatism.

If the incident beam is parallel (see Fig. 3.6) then the diffracted beams leaving the specimen are brought to a focus in the back focal plane of the objective lens and this is where the diffraction pattern is first formed. In order to observe a diffraction pattern (Fig. 3.6b) the back focal plane of the objective lens must be projected onto the viewing screen and this is achieved by running P1 at reduced

48 THE ELECTRON MICROSCOPE

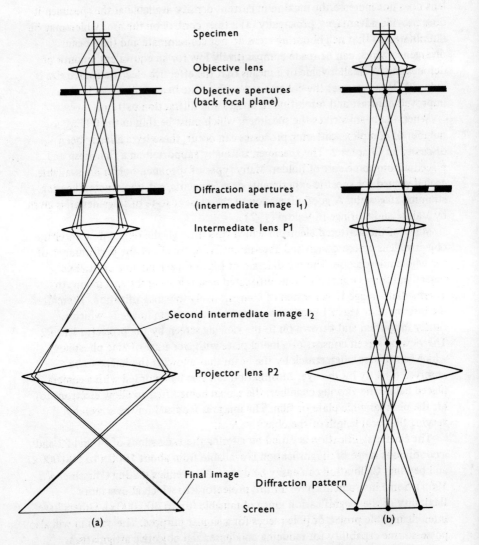

Figure 3.6 Ray diagrams in the CTEM. (a) Imaging conditions. (b) Diffraction conditions. In (a) the I_1 plane and the final screen are conjugate. In (b) the back focal plane of the objective lens and the final screen are conjugate.

A BRIEF PHYSICAL DESCRIPTION OF THE CTEM 49

excitation. A series of interchangeable objective apertures varying in size from 50 μm to 200 μm is placed near the back focal plane of the objective lens which have an important role in the production of contrast. In the deficiency contrast mode the objective aperture is placed to intercept all the diffracted beams and allows only the direct or undeviated beam to pass through, thus forming what is called a bright field image (Fig. 3.7). It is evident that contrast in the image is produced by differences in the intensities of electrons scattered coherently from various parts of the specimen: hence the term diffraction contrast. Even with non-crystalline specimens, such as glasses and replicas, the objective aperture is inserted to restrict some of the scattered electrons from contributing to the final image. Its insertion also reduces spherical aberration.

Fig. 3.7b illustrates diagrammatically a view of the back focal plane of the objective lens. Diffraction spots can be seen with the objective aperture placed round the undeviated beam so as to produce diffraction contrast conditions. Normally the aperture is placed to isolate the direct beam but imaging of a particular set of lattice planes in a crystalline object may be accomplished (see Fig. 4.22) if the beams diffracted by these planes together with the direct beam are allowed to pass through the aperture. The image produced may be considered as a type of interference pattern. This technique is rather specialized and has its widest application as a test of resolution of the microscope. It is discussed more fully in Chapter 4.

The spacing of the diffraction spots (or rings if the specimen is polycrystalline) at the back focal plane depends upon the excitation of the objective lens. In some machines a facility is available for observing the specimen at a position raised above the 'standard' position. In this case the lens excitation is weakened, the back focal plane displaced (it no longer coincides with the objective aperture) and the spacing of the spots increases, i.e. the 'magnification' of the pattern is increased.

A counterpart to the bright field mode of operation exists and is called dark field imaging. This time the objective aperture is used to isolate a single chosen diffracted beam rather than the undeviated beam. This can be done by displacing the objective aperture but astigmatism is produced in the image. A more satisfactory method is to tilt the illumination as shown in Fig. 3.7c. In crystalline studies it is wise to take both dark and bright field micrographs in order to extract the maximum information and examples of the technique will be found in Chapter 4.

A most important facility of the CTEM makes use of the intermediate or diffraction apertures placed in the I_1 image plane (see Fig. 3.6b). If an aperture of diameter D is positioned in this plane only electrons passing through a circle of size D/M (M the objective magnification) of the specimen will reach the

50 THE ELECTRON MICROSCOPE

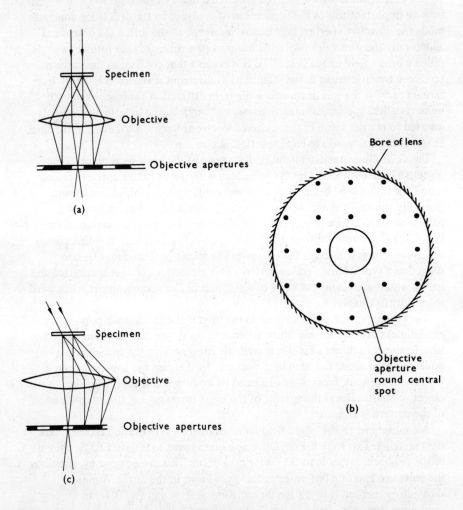

Figure 3.7 Bright and dark field imaging. In bright field (a), (b), the objective aperture is placed to allow only the direct beam to form the final image. For dark field imaging (c) the illumination is tilted to allow through a chosen diffracted beam.

screen. If $M = 10\times$ and $D = 10$ μm, the diameter of the specimen area viewed will be ~ 1 μm. A diffraction pattern from this restricted area can also be observed. This technique is known as selected area diffraction and enables a correlation between micrograph and diffraction features to be made on a very fine scale. For correct operation of this procedure lens P1 should be focussed on the I_1 plane (i.e. on the aperture) an excitation often marked on the instrument, and then the specimen focussed with the objective. When using the diffraction facility for the determination of lattice parameters, the best accuracy (about $0\cdot1\%$) is achieved with small values of illumination angle at the specimen. The dimensions of the diffraction pattern on the viewing screen depend upon the camera length defined as the product of the objective focal length and the total magnification of P1 and P2 when operating in the diffraction mode. This length is therefore a function of the lens excitations and calibration should be undertaken with a standard specimen e.g. a film of evaporated gold or thallium chloride. For improvement of resolution many instruments are now equipped with ancillary high resolution diffraction stages which are inserted either between the P1 and P2 lenses or below them both, depending upon the particular design.

Sometimes the individual diffraction spots carry structural or magnetic information which it is desired to study. One way of doing this is to increase the camera length — up to several hundred metres if necessary — so that one spot may occupy a substantial area of the viewing screen. Various 'low angle diffraction' modes which achieve this end by modifying the conventional lens excitations have been devised. For details of these specialized techniques the reader is referred to Ferrier (1969).

3.5 A brief physical description of the SEM

A schematic diagram of the electron optical column of the SEM is given in Fig. 3.8. Although two lenses are depicted this number is by no means standard. The electron gun works in the range 0–50 kV (the latter limit again depends upon the make of machine) while the specimen is maintained at earth potential. The microscope is essentially a probe forming instrument, each of the lenses performing a condensing action, successively demagnifying the beam source (size $d_0 \sim 50$ μm) down to a focussed spot size d at the specimen. The relation between d and d_0 is

$$d = M_1 M_2 d_0 \tag{3.10}$$

where M_1 and M_2 are the demagnifications of the lenses. Strictly speaking Equ. 3.10 is only valid in the absence of aberrations when d becomes the Gaussian

52   THE ELECTRON MICROSCOPE

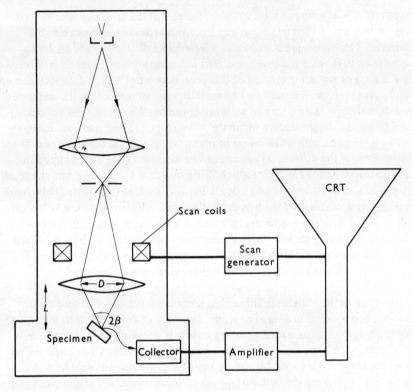

Figure 3.8   A schematic diagram of the scanning electron microscope.

spot size. In actuality because aberrations do exist the spot size on the specimen, d_f, is somewhat larger than d; this will become clearer in the next section. M_1 and M_2 are controlled by the lens excitation currents and the latter also depends upon the working distance L. The electron beam falling on the specimen is often termed the 'probe' and carries a current typically of 100 pA.

Despite its demagnifying action the final lens is sometimes referred to as the objective, possibly because it is used to focus the specimen. The working distance L, and the final lens aperture diameter D, have a strong control on the divergence of the beam which in turn determines the current density at the specimen. Commonly found values for D are 50, 100 and 200 μm and for L, which can be varied, about 10 mm; β will be about 10^{-2} rad.

The microscope gets its name from the action of two sets of scanning coils incorporated in the vicinity of the final lens. In the most common arrangement, when energized by a suitable scan generator, the coils cause the beam to pivot

A BRIEF PHYSICAL DESCRIPTION OF THE SEM   53

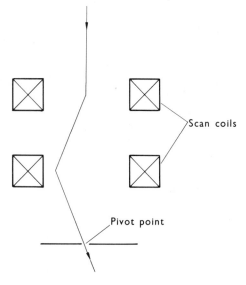

Figure 3.9 A double deflection scan system which pivots the electron beam about the centre of the final lens aperture. Only the coils producing deflections in the plane of the paper are shown.

about the centre of the final lens aperture (Fig. 3.9). The beam is thus made to deflect over the specimen in the form of a raster. The raster action is similar to that which takes place in a television tube where the electron beam sweeps out the screen area in 625 lines. Another useful facility desirable in certain applications is that of beam 'chopping' i.e. the ability to turn the beam on and off in a controlled way over a range of frequencies. Chopping stages are usually inserted near the filament of the microscope.

Many of the physical effects that occur when the probe strikes the specimen surface at a given point give rise to signals which can be collected and amplified to give information about that point. Each signal can be collected simultaneously and passed to the control grids of different display cathode ray tubes which are scanned synchronously with the electron beam in the main column. In this way the same square raster is built up on the face of a display tube and the brightness variations seen are due to signal variations from point to point on the specimen surface. For the sake of simplicity only one collector and amplifier system is shown in the figure. Every instrument has at least two cathode ray tubes; one for video display with a long persistence time, the other for photographic recording. In addition to these normal capabilities there will probably be

54   THE ELECTRON MICROSCOPE

facilities for the display of a single line scan taken at will somewhere across the specimen and also Y−modulated signals. In Y−modulation a raster is built up of line scans which have constant brightness but 'plot out' the intensity variations.

For most purposes the standard operating mode is the emissive mode in which mainly secondary electrons are collected. A simple collector for secondary electrons of a type due initially to Everhart and Thornley (1960) is illustrated in Fig. 3.10. Electrons are attracted from the specimen at earth potential towards a metal cup with a metal grid over its opening held at about + 250 V. Any electrons which penetrate the grid are accelerated through a potential difference of about 10 kV towards a plastic scintillator covered with a thin film of aluminium. The light created in the scintillator passes down a light pipe to a photomultiplier where it is converted to an electric current ready for subsequent amplification. The flight time of the electrons is very short − of the order of 10^{-7} s. This device can also be used for detecting back scattered primaries, if the potential on the grid is made slightly negative to discourage the collection of low energy secondaries. Moreover, by removing the scintillator and light pipe entirely, cathodoluminescent radiation can be made to fall directly on the photomultiplier; the collection of light can be improved by using a simple lens or mirror. A further simple modification to the basic design permits the enhance-

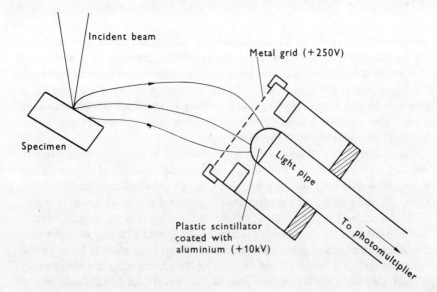

Figure 3.10   A schematic representation of a collector suitable for secondary and reflected electrons.

A BRIEF PHYSICAL DESCRIPTION OF THE SEM 55

ment of magnetic and electric field contrast (see Section 5.2.2). Notwithstanding the versatility of the Everhart—Thornley collector others have been developed, in particular the solid state or semiconductor detector. Here the incident electrons strike the detector and produce electron—hole pairs which cause the flow of an electric current in an external circuit. Since detection is achieved electronically — no image in the usual optical sense being formed — a great deal of signal processing is possible in the SEM, far more so than with the CTEM. Various types of signals may be added or subtracted to suit particular requirements. An account of detector systems is given by Wells (1974).

The magnification of the microscope is controlled by the scan generator which varies the ratio of the raster size on the specimen to that on the video cathode ray tube, the latter normally being 100 mm x 100 mm. If the generator supplies voltage to make the specimen raster 10 μm x 10 μm the effective linear magnification is 10^4. Obviously as the magnification is increased the angle of scan which the beam makes over the specimen is progressively diminished. On standard machines a magnification range of perhaps 20—50 000x is available. To obtain the lowest magnifications the specimen must be lowered, thereby increasing the working distance and making spherical aberration more severe. Besides vertical movement, the stub, on which the specimen is mounted, can be rotated or tilted with respect to the incident beam, a procedure which increases the signal (see Section 2.3.1). If the specimen is tilted fore-shortening may occur and this should be allowed for when estimating magnifications.

A relationship exists between the useful magnification and the probe size at the specimen. This may be seen as follows (Fig. 3.11). Assuming that the resolution of the eye is 0·1 mm and the viewing raster has size l mm x l mm (l is

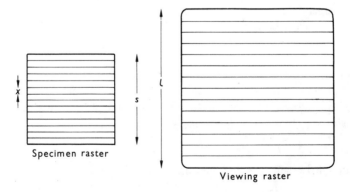

Figure 3.11 The relationship between the raster size on the specimen and the viewing raster on the video CRT.

usually 100) the number of raster lines which may be usefully displayed on the cathode ray tube is $l/0\cdot 1$. If the corresponding raster size on the specimen is s, the linear magnification $M = l/s$. The interval between line scans on the specimen is given by $x = 0\cdot 1 s/l = 0\cdot 1/M$, i.e. x is a function of M. Clearly the probe size, d_f, should be equal to x because if not, either the lines overlap $(d_f > x)$ or certain parts of the specimen are not scanned at all $(d_f < x)$. For $M = 1000\times$, the focussed spot size should be 100 nm, and for $M = 10\ 000\times$, 10nm. The value of d_f is a strong function of the lens excitations, Equ. 3.10, which should be adjusted accordingly. Failure to fulfil the condition $d_f = x$ may give rise to misleading information. The simple treatment outlined here, though instructive, assumes the probe to be uniformly bright. In fact the intensity falls at the edges so that a more realistic condition is for the scan lines to be somewhat closer together than d_f.

In common with the conventional transmission microscope, the SEM has a very large depth of focus, (up to several millimetres). Undoubtedly this property is one of its most valuable assets.

3.6 Considerations of resolution

The resolution obtainable in an electron microscope is a quantity of paramount importance and in this section are considered some of the factors which affect the resolution in the two principal types of microscope.

3.6.1 The CTEM

To obtain a theoretical value for the ultimate resolution of a microscope is no easy task, depending as it does upon the coherence of the electron beam (itself a function of the electron gun), the nature of the object and the aberrations of the lenses. Nevertheless a reasonable estimate of the resolving power can be found by considering the spherical aberration of the objective lens together with diffraction effects. The spherical aberration of the remaining lenses can be neglected because of the small angular beam divergences involved.

From Equ. 3.8 the radius of the aberrant disc at the object — neglecting for the moment any chromatic defect — is given by

$$r_{aber} = (r_s^2 + r_d^2)^{1/2} \tag{3.11}$$

Substituting for r_s and r_d from Equations 3.5 and 3.7, then

$$r_{aber} = \left[\left(\frac{0.61\lambda}{\alpha}\right)^2 + C_s^2 \alpha^6\right]^{1/2} \tag{3.12}$$

CONSIDERATIONS OF RESOLUTION   57

By solving the equation $dr_{aber}/d\alpha = 0$, a minimum value for α is obtained which when substituted back into Equ. 3.12 yields a value for the resolution r_{min}. Performing these operations gives

$$\alpha = 0.77[\lambda/C_s]^{1/4} \qquad (3.13a)$$

and

$$r_{min} = 0.91\ C_s^{1/4}\lambda^{3/4} \qquad (3.13b)$$

For an objective working at 100 kV, $\lambda \simeq 4$ pm and a typical value of C_s is 3 mm; from Equ. 3.13 the optimum divergence angle would be $\sim 4 \cdot 5 \times 10^{-3}$ rad and the limit of resolution $\sim 0 \cdot 6$ nm. Several remarks should be made about this result. It is apparent that the resolution is much inferior to that which would obtain if it were limited by diffraction only, as is generally the case in the light microscope. Second, the resolution predicted is only a guideline and does not necessarily mean that object points or lattice planes with a separation distance less than $0 \cdot 6$ nm would not be observed. To begin with it is likely that some improvement in resolution will result if the specimen is slightly defocussed since the size of the disc due to spherical aberration will diminish It is a practice of some manufacturers to quote both a point-to-point resolution and a lattice plane resolution, the latter being somewhat the lower figure. Finally other simple ways can be used for combining aberrations than the one adopted here. Thus the sum $r_s + r_d$ could be minimized with respect to α or even the equality $r_{min} = r_s = r_d$ could be solved. However a result of the form $r_{min} =$ constant. $C_s^{1/4} \lambda^{3/4}$ is still obtained.

Equ. 3.13b implies that resolving power can only be improved by reducing C_s or λ. The latter possibility is offered by going to increased accelerating voltages but the benefit is unfortunately offset by increasing values of C_s associated with the lenses required.

For certain types of specimen it may be chromatic aberration that imposes the final limit to resolution rather than spherical aberration. This is because fast electrons incident upon a specimen lose energy through scattering processes of various kinds (see Chapter 2) and so will not be correctly imaged by the objective lens. In a thick sample these losses may amount, say, to 10 eV. Thus for an accelerating voltage of 100 kV, an objective with a chromatic aberration constant $C_c = 2$ mm and $\alpha = 5 \times 10^{-3}$ rad, the size of the aberrant disc (Equ. 3.6) is $r_c = C_c \alpha \Delta V/V = 1$ nm, clearly a not inconsiderable value.

No attention has been paid to the harmful effects of other aberrations, such as astigmatism or distortion, since adequate means for their control are incorporated in the microscope column. Any failure to do so will naturally reduce performance. Eventually of course the final image will be projected on to a fluorescent screen or photographic plate, whereupon a limitation imposed by the resolving power of the human eye (0·1 mm) must be considered. If the limit of resolution of the microscope is r_{min}, the maximum useful magnification on the screen or plate is $0·1/r_{min}$(mm). Thus for $r_{min} = 0·4$ nm, the highest useful magnification is 250 000 x; greater magnification will yield no more detail.

3.6.2 The SEM

The resolution of the scanning microscope can be defined as the smallest extent between two features on the specimen from which signals can be produced and discerned. It appears that resolution is therefore essentially a question of probe size at the specimen and signal-to-noise ratio. However other factors can intrude, e.g. mechanical vibration or stray magnetic or electric fields and more importantly, the spreading effects produced inside the specimen. The latter have already been described in Chapter 2 and give rise to the familiar 'tear drop' distribution. The extent of the spreading depends upon the operating mode (e.g. X-ray or voltage contrast), the accelerating voltage and the type of specimen: obviously some of these are beyond control in a particular situation.

It is clear that the resolution can never be less than the size of the probe at the specimen. At this juncture the aberrations of the final lens must be considered because their effect is to increase the Gaussian spot size d which, it will be recalled, is defined as the probe size in the absence of aberrations. Some distinctions should be noticed concerning the treatment of aberrations for the two principal types of microscope. In the CTEM we are concerned with their degrading influence in object space (i.e. at the specimen) whereas for the SEM it is the image plane of the final lens which is of prime interest. (The fact that the image of the source produced by this lens coincides with the specimen can be a little confusing.) Thus it is more convenient in the latter case to express the aberrations as a function of β, the beam divergence at the specimen rather than α as before. Conventionally also, the diameters of the aberrant discs are considered, not the radii. Finally the objective lens of the CTEM and the final lens of the SEM play inverse roles; the function of the former is to magnify while the latter demagnifies. For the conditions of operation of a scanning microscope the diameters of the discs of confusion can be expressed in the form (Oatley, 1972)

$$d_d = 1.22\lambda/\beta; \quad d_s = \tfrac{1}{2}C_s\beta^3; \quad d_c = C_c \frac{\Delta V}{V} \beta \qquad (3.14)$$

Note the similarity with Equations 3.5, 3.6, and 3.7. d_s as presented here represents the diameter of the disc of least confusion. The aberration constants C_s and C_c found in Equ. 3.14 are the same as those that would be appropriate if the lens were used as an objective. This is because in both cases we are interested in aberration effects near a focal plane of the lens. For more details on the relation between the aberration constants for magnifying and demagnifying lenses the interested reader is referred to Hawkes (1972) and Oatley (1972). Typical values for the final lens in the SEM are C_s = 20 mm and C_c = 8 mm. Using an equation analogous to 3.8, the size of the disc due to aberrations, d_{aber}, can be obtained and is in the region of 5 nm.

At this point it is pertinent to mention a recent development due to Wells *et al.* (1973). Compared with the objective lens of the CTEM, the final lens of a 'normal' scanning microscope has a long focal length and therefore worse aberration characteristics. In the device of Wells *et al.*, the specimen is actually located inside the final lens which has a short focal length with minimal aberrations. The lens is so designed as to deflect away secondaries and the signal, contrary to the usual state of affairs, is comprised of reflected primaries.

Leaving aside the problem of aberrations, it serves no useful purpose to reduce the probe size to a small value if the number of electrons striking the specimen does not generate a large enough signal to overcome noise. The noise may arise from random fluctuations in the incident electron beam, from scattering processes within the specimen or from amplification and collection devices.

The amount of noise tolerated depends on the degree of contrast required in the image. Fig. 3.12 illustrates image features with contrast $C = \Delta I/I$ in the presence of a uniform noise level ΔN. Clearly the perception of contrast against a noise background is to some extent subjective. A commonly adopted criterion in scanning microscopy is that to discern a contrast feature ΔI against a background I in the presence of a superposed noise level ΔN, then $C \geqslant 5\Delta N/I$. If the number of electrons incident upon a particular element of the surface is n, the statistical fluctuation (noise) associated with this number is $n^{1/2}$, i.e. $\Delta N/I = n^{1/2}/n$. The signal-to-noise criterion thus becomes $n \geqslant 25\, C^{-2}$. To take account of noise arising from other sources this can be amended to

$$n \geqslant 100 C^{-2} \tag{3.15}$$

This formulation may not be rigorous but it does illustrate the important notion that to detect finer shades of contrast n must be made larger.

It is common practice when discussing noise considerations to use the idea of 'picture point'. If a complete raster consists of N (often 10^3) lines per frame, each line comprises N picture points and the whole frame contains N^2 picture points. In other words the total raster area is divided into N^2 elements each of diameter equal to the Gaussian spot size. According to Equ. 3.15 if a certain contrast level is required a definite number, n, of electrons must fall upon each

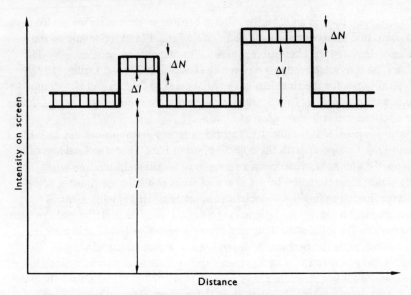

Figure 3.12 The superposition of a constant noise level ΔN on signals ΔI against a background intensity I. The intensity distribution is schematic only.

picture point, e.g. a contrast of $C = 5\%$ demands that $n \geqslant 4 \times 10^4$. Various values of absolute currents in the probe, $\mathscr{I}$, and scanning times, t, (the time to complete one raster) combine to give the same value of n. The actual relationship between these quantities is

$$n = \mathscr{I} t / e N^2 \qquad (3.16)$$

Suitable combinations to achieve $n = 4 \times 10^4$ electrons per picture point assuming a value of $N = 10^3$ are; 5 nA in one second or 0·5 nA in 10 seconds etc. $\mathscr{I}$ may be related to the Gaussian probe size and fundamental constants of the microscope gun with the Langmuir formula, Equ. 3.3, in which case combining Equations 3.15 and 3.16 we obtain

$$\frac{\pi i_0 V t d^2 \beta^2}{N^2 k T_0} \geqslant 400 C^{-2} \qquad (3.17)$$

Equ. 3.17 expresses an adequate signal-to-noise ratio criterion in terms of the microscope variables normally available. Thus all other things being equal, as d is reduced the concomitant loss of signal must be compensated by an increase in β. It should be emphasized that Equ. 3.17 is meant only as a guideline.

To arrive at an overall figure of best resolution it remains now to re-introduce the deleterious effects of aberrations. The final probe size d_f, in the presence of aberrations will be given by

$$d_f^2 = d^2 + d_{aber}^2 = d^2 + d_d^2 + d_s^2 + d_c^2 \qquad (3.18)$$

where d is the Gaussian probe size derived from Equ. 3.17 and d_d etc. are the aberrant disc sizes obtained from Equ. 3.14. Each term in the above equation is a function of β which can be optimized to give the minimum value of d_f and thus an estimate of the best resolution. A good value of d_f for a modern instrument is 10 nm although this can be expected to improve in the future with the introduction of such features as higher brightness electron sources. However, even the figure of 10 nm should be accepted with qualifications. No quantitative account has been taken above of the spreading of the beam inside the specimen; in certain instances this can dominate the resolution. One way of reducing the spreading is by lowering the accelerating voltage but unfortunately this also increases d_{aber} and decreases the signal-to-noise ratio. It should also be borne in mind that the best possible resolution is useful for one setting of magnification only and that fixing a value of β sets the depth of focus (Equ. 3.9). In many applications it might be preferable to sacrifice resolution for increased depth of focus.

In conclusion, the performance of the SEM is seen to depend upon a number of inter-related parameters. The account above is meant only as an introduction and for further details of practical operation more specialized literature should be consulted, e.g. Booker in Amelinckx et al. (1970) and Oatley (1972). In addition the handbook supplied with an instrument will give a graphical guide to many of these factors and so help to optimize results.

3.7 The merits and disadvantages of electron microscopes

The most important advantage of electron microscopes is, naturally enough, their excellent resolution when compared with that of the optical microscope. Vastly increased depth of focus in the SEM is also of especial value. Nevertheless there is a price to be paid besides the obvious financial one: this is mainly the suitability and preparation of specimens. These restrictions are more severe in the CTEM where thin sections or replicas must be produced, often at the expense of much time and effort. The SEM accepts much bigger specimens and also offers more manoeuvrability with regard to their positioning. This is a useful facility when dynamic experiments are contemplated. There are particular difficulties associated with the observation of living biological organisms which barely survive — if at all — exposure to high energy incident electron beams: even dead biological material may suffer. Despite these drawbacks it is hoped that the applications presented later will prove the efficacy of electron microscopy and also serve to compare and contrast the capabilities of the instruments.

4

Conventional Transmission Electron Microscopy

4.1 Introduction

Before going into details of how transmission electron microscopy can be used to help in the determination of the structure of materials it is useful to discuss in a general manner the area of interest. In most cases the CTEM can be used to derive information of several different kinds which extend right across the sciences concerned with elucidating microstructure. The external surface of a body can be studied and information obtained concerning the external morphology of the specimen and also microscopic details of the surface roughness can be investigated. Materials of interest here are fibres and small particles in which the natural surface has a direct bearing on the properties and uses of the material. However, even in materials in which surface properties are not important much information about the constitution of the material can be achieved by studying a prepared surface. The CTEM may be used either to cast a shadow of the specimen, if it is sufficiently small, or to observe a replica of the specimen surface. Information about the internal structure of a material is obtained directly in transmission and, in most cases, the thickness of the specimen has to be deliberately reduced to obtain a condition of transparency for electrons of energy in the range 100–1000 keV.

In this chapter some of the above aspects will be considered at some length in certain particular cases, without necessarily describing in detail the techniques used to prepare the specimens. References in this latter area together with further reading in the aspects to be covered here will be given at the end of the book.

4.2 Surface information and external morphology

Perhaps one of the first applications of the CTEM was to study the size, shape and dispersion of small particles. Here, of course, transmission of electrons is not essential and the microscope is used as a super optical microscope of great magnifying power. Shadow micrographs of particles and fibres show the shape of the object and, if the specimen preparation is well controlled, a typical state of dispersion of the particles. Important applications in this area are the study of particle shape and size distributions (a) from solution such as colloidal preparations, soil fractions and precipitates and (b) from 'dry' origins such as airborne dusts, paint pigments, powders and fibres. If the specimen has a characteristic shape, such as that possessed by certain minerals and viruses, a study by electron microscopy can be an aid to identification. An added bonus of crystal structure identification by electron diffraction is obtained if crystalline particles occur naturally as thin platelets or are produced in this form.

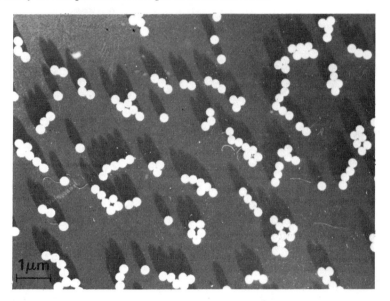

Figure 4.1 A transmission electron micrograph of shadowed latex particles supported on a thin carbon film (courtesy of P. Chippindale).

Two examples incorporating shadow microscopy to show particle shape and size distribution and possibilities for electron transmission and diffraction are given in Figs 4.1 and 4.2. The first is of a distribution of latex particles supported on a thin carbon film. The spherical form of the particles is clearly indicated by their circular image and the 'shadow' cast to one side of each particle. The technique of shadowing is considered later but it is obvious from this example that it can be used to give an impression of the 'height' of a feature and its shape. The uniformity in the particle diameter is also illustrated.

64 CONVENTIONAL TRANSMISSION ELECTRON MICROSCOPY

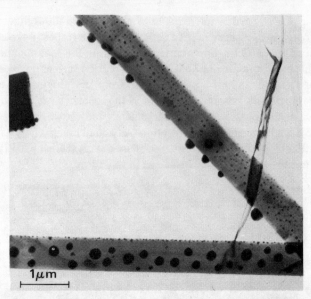

Figure 4.2 Silicon nitride fibres coated with nickel. The nickel coating has been spheroidized into small particles at high temperature. (Courtesy of E. H. Andrews *et al*, *J. Mater. Sci.*, 1972, **7**, 1003.)

An area of interest in materials science is illustrated in Fig. 4.2 where the action of temperature has spheroidized a nickel coating on some Si_3N_4 fibres; the shape of the nickel particles is clearly shown and on some the beginning of faceting, i.e. the formation of identifiable crystal facets, is evident. In this example the potentialities of electron diffraction for identifying crystal orientation and relationships at the fibre/coating interface could be used, in principle at least, as some transmission of electrons through the fibres is evident. In some places in the fibres, features attributable to diffraction contrast (see Chapter 2) are present.

In the examination of electrically insulating fibres or particles (or indeed any insulating specimen) a conducting coating must sometimes be applied to the specimen. However, if a supporting carbon film is used this can sometimes be avoided and charging effects can also be reduced by working at a reduced accelerating voltage and beam brightness. This is often convenient as transmission of electrons is not always required.

The CTEM has to a large extent been superseded in this area by the SEM except when high resolution is required. Indeed, as a routine application and where a three-dimensional morphology is sought the scanning mode is by far the most efficacious technique. This is true in many problems which some years ago

SURFACE INFORMATION AND EXTERNAL MORPHOLOGY 65

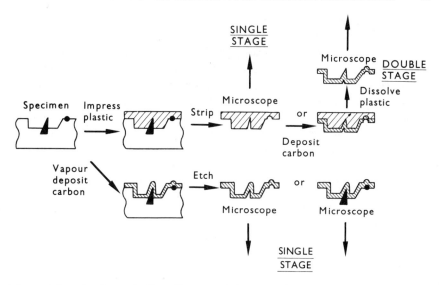

Figure 4.3 A schematic flow diagram of the stages in the preparation of plain and extraction replicas for electron microscopy.

(before ~1966) would have been investigated by replication methods associated with transmission electron microscopy. The development of the double condenser system also narrowed the range of usefulness of the replica in that metallic and inorganic materials in a sufficiently thin form could be examined in transmission. However, where high resolution is concerned and surface detail of the order of 2–5 nm is of interest replication methods are still widely used and indeed may be the only ones available for examination at resolutions significantly better than that of the optical microscope for some time to come.

Replicas are indirect specimens which replicate the surface structure of interest. They may be classed as either single or double stage according to their method of fabrication as illustrated in Fig. 4.3 which shows schematically a prepared surface containing an exposed second phase. This represents a general situation in that not only does the specimen have a surface structure but it is also a composite material and some of the included material has been unknowingly or deliberately exposed. The two single stage methods involve replication by an evaporated (vacuum deposited) carbon film or by an impressed plastic film. Contrast is obtained in the CTEM through the difference in specimen thickness at the indentations and inclinations in the replica film and explained by mass thickness concepts as discussed in Chapter 2. The inclusion of the second phase in the replica is an important technique known as extraction replication and can be used to identify precipitates or included particles in alloys using electron diffraction. The two stage carbon replica gives improved resolution (~2 nm) over the single stage plastic film but the single stage replica is popular because of the ease and speed of its fabrication. Carbon replication has its merits in good resolution, i.e. faithful reproduction of detail, strength and stability in the electron beam. Two stage methods are required in carbon replication if it is necessary to keep the specimen surface in its original condition. Contrast in replication can be improved by 'shadow casting' the film with a heavy metal such as a Au/Pd alloy or Pt/C mixture as illustrated in Fig. 4.4. The heavy atoms scatter electrons at appreciable angles and a considerable fraction is stopped by the objective aperture. The shading obtained gives a quantitative estimate of the topography of the original surface but care must be taken to check whether the replica is a positive or negative of the original surface.

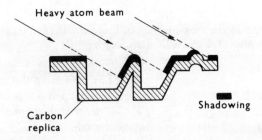

Figure 4.4 Illustrating the action of shadowing to increase contrast and reveal surface features in replicas.

Two examples of replication from widely different problems in materials science are shown in Fig. 4.5. The first is a Pt/C coated replica of the cross section of a multilayer thin film optical filter consisting of a series of dielectric and metal films evaporated sequentially in vacuo on to a glass substrate. The thickness and granular structure of each layer is clearly visible and defects in the layered structure at several positions are clearly revealed. Structural information of this kind is at, or beyond, the present limit of resolution of the SEM and could not be obtained by any other generally available microscopic technique. The second example is of a carbon extraction replica containing extracted precipitates of TiO_2 from an internally oxidized Cu–Ti alloy. The diffraction pattern characterizing the TiO_2 is shown in the inset. Thus, not only has

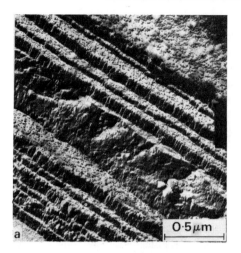

Figure 4.5 Examples of replicas (a) a Pt/C coated replica of the cross section of a multilayer thin film optical filter (Courtesy of J. Pearson) and (b) an extraction replica of TiO_2 particles and their diffraction pattern (Courtesy of G. R. Woolhouse).

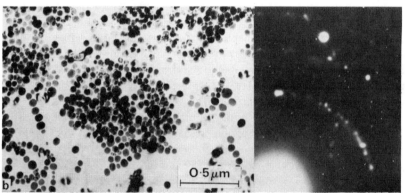

information about the size distribution and shape of the precipitates been obtained, but their crystal structure has been elucidated. However, it should be remembered that the chemical nature of the precipitates has not been analysed, only the crystal structure and 'd' spacings which agree with those already known for TiO_2 have been obtained. New 'hybrid' electron microscopes capable of chemically analysing materials are discussed in Chapters 5 and 6. Attempts to obtain diffraction patterns from the titania, whilst still in the parent thin section of copper, were frustrated by the low concentration of precipitates giving relatively low intensities, and complications in diffraction at the interface of the copper and the titania. An important point to note about extraction methods is that the dispersion of the second phase particles that occurs in the original

68 CONVENTIONAL TRANSMISSION ELECTRON MICROSCOPY

matrix is not retained in the replica and any orientation relationship between the matrix and the extracted particles is also lost.

The details of specimen preparation and 'tricks of the trade' for the areas briefly covered in this section are many and often highly specialized. The fundamental details are dealt with by Brammar and Dewey (1966) for metallography and by Kay (1965) for fibres and particles and associated materials. The latter work gives many particular examples and a recent contribution by Goodhew in Glauert (1972) gives an exhaustive reference list and details of modern developments in techniques related to materials science.

4.3 Electron diffraction and simplified theories of amplitude or deficiency contrast in crystals

It is probably true to say that most of the contrast effects studied in detail in transmission electron microscopy originate in crystalline materials. Non-crystalline materials such as the replicas discussed above, glasses and polymers produce contrast that is usually easily interpreted, qualitatively at least, in terms of mass thickness concepts (see Chapter 2). Some special contrast effects in this class of materials are discussed briefly in Section 4.5. Crystals give a wealth of contrast detail that is intimately related to the structure of the material and its departure from perfection, be it due to the presence of lattice faults such as dislocations or to second phase particles. This contrast can be explained in a real quantitative sense using calculations of the intensity of electrons diffracted into particular reflections in a given situation. Because of the success of these theories and the wide range of materials to which they apply it is necessary to consider them in a little detail although more extensive and advanced treatments are given by e.g. Heidenreich (1964), Hirsch et al. (1965) and contributions from Brown, Gevers and Howie in Valdré (1971).

Diffracted intensities can be rigorously calculated using a theory of diffraction called the dynamical theory of electron diffraction. In this approach multiple scattering is allowed for, i.e. the diffracted intensities are large and the diffracted waves can themselves be scattered. Dynamical events can therefore occur where the incident and scattered waves interact; such interactions occur also in X-ray diffraction of course and dynamical theories of them exist. One scheme that can be used to solve for the scattered intensities is a wave mechanical treatment where the crystal is treated as a periodic electrostatic potential and the electron wave functions (known as Bloch wave functions) are solutions of a steady state Schrödinger equation. Here the incident electron wave is scattered from the potential field or the crystal as a whole, rather than from individual atoms. This kind of approach will be familiar to those who have studied the energy band theories of solids see, e.g. Coles and Caplin (1975) and Dugdale (1975) where the propagation of electron waves through a periodic crystal lattice potential is discussed. Detailed treatments of dynamical theories are beyond the aim of this book and are not necessary for a qualitative description of diffraction contrast. However, it is true to say that certain observed effects are only explained by the dynamical theory and for this reason a brief discussion of it is given here.

The Bloch wave functions $\Psi(r)$ in the periodic lattice *potential* $V(r)$ are a solution of the Schrödinger equation

$$\nabla^2 \Psi(r) + \frac{8\pi^2 me}{h^2} [E - V(r)] \Psi(r) = 0 \qquad (4.1)$$

Here, the total electron energy is eE and is defined by Equ. 4.3 below.

In a 'two beam' approximation when only one diffracted beam of wave number $K + g$ is produced, substitution of $\Psi(r)$ and $V(r)$ in their complete forms into Equ. 4.1 results in a relation of the form

$$(K - K_r)(|K + g| - K_r) = (2me V_g/K_r h^2)^2 \qquad (4.2)$$

(see e.g. Hirsch *et al.* (1965)). K_r is the incident wave after the correction for refraction by the mean inner lattice potential V_0 on entering the crystal (V_g are the Fourier coefficients of the potential) and is related to the wave number in vacuum, χ_v, as

$$K_r^2 + \left(\frac{2me}{h^2}\right) V_0 = \frac{2meE}{h^2} = \chi_v^2 \qquad (4.3)$$

Equ. 4.2 relates the wave number of the electron to its energy, as does Equ. 4.3, and defines a dispersion surface. (In a metal the Fermi surface is a dispersion surface for electrons of energy equal to the Fermi energy). In this case it has two branches, asymptotic to the spherical surfaces defining K_r which have origins at O and P as shown in Fig. 4.6. At the position in the reciprocal lattice where the Bragg condition is satisfied, which is coincident with the Brillouin zone boundary, the dispersion surfaces are separated by $\Delta K = (2me V_g/K_r h^2)$ and hence K and $K + g$ must each have two values $K^{(1)}$, $K^{(2)}$, $K^{(1)} + g$ and $K^{(2)} + g$. The incident and reflected waves combine to give two standing waves

$$\Psi^{(1)} = \exp 2\pi i(K^{(1)} \cdot r) - \exp 2\pi i[(K^{(1)} + g) \cdot r]$$
$$\Psi^{(2)} = \exp 2\pi i(K^{(2)} \cdot r) - \exp 2\pi i[(K^{(2)} + g) \cdot r] \qquad (4.4)$$

which are standing waves with respect to the direction perpendicular to the lattice planes giving the Bragg reflection. The electron current (corresponding to

70 CONVENTIONAL TRANSMISSION ELECTRON MICROSCOPY

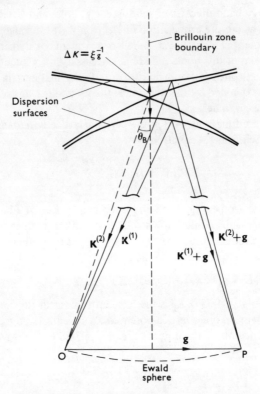

Figure 4.6 Showing the dispersion surface in the dynamical theory and the excitation of waves $K^{(1)}$, $K^{(1)} + g$, $K^{(2)}$ and $K^{(2)} + g$. The separation of the surface branches at the exact Bragg condition is $\Delta K = \xi_g^{-1}$.

the component of K parallel to the lattice planes) will have an intensity which will be modulated along a perpendicular to the planes such that maxima in the intensity corresponding to wave (1) occur midway between the planes and those for wave (2) are concentrated at the planes.

The modulation can be understood in a simple way by considering plane waves travelling in a one dimensional lattice of parameter a. At the value of K_x corresponding to the Bragg condition, $K_x = \pm 1/2a$, the incident and reflected plane waves $\exp(i\pi x/a)$ and $\exp(-i\pi x/a)$ combine on addition and subtraction to give two standing waves $\Psi^{(1)} = \sin \pi x/a$ and $\Psi^{(2)} = \cos \pi x/a$. Hence the charge density $e\Psi^2$ corresponding to waves with x components of their wave vectors equal to $1/2a$ and hence the electric current parallel to the planes is peaked for $\Psi^{(1)}$ at $x = a/2$ and for $\Psi^{(2)}$ at $x = a$. This is illustrated schematically in

ELECTRON DIFFRACTION 71

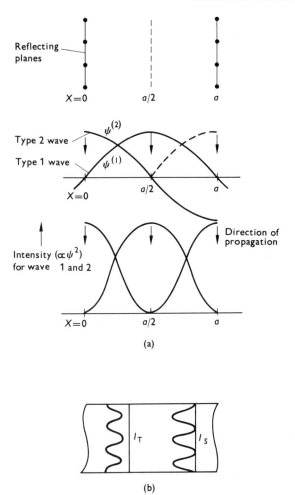

Figure 4.7 A schematic representation of (a) the amplitude in the type 1 and 2 waves peaked in between and at the atomic planes and (b) the oscillations with depth of the transmitted and scattered intensities in the dynamical theory.

Fig. 4.7(a). Since wave (2) concentrates charge in a region of lower potential energy in the lattice than wave (1) it has greater kinetic energy and hence $K^{(2)} > K^{(1)}$. This difference in wave number leads to a beating effect between $\Psi^{(1)}$ and $\Psi^{(2)}$ and to oscillations in the transmitted intensity (I_T) and the diffracted intensity (I_s) as indicated in Fig. 4.7(b). At the exact Bragg position the periodicity of the oscillations is known as the extinction distance $\xi_g = (2meV_g/K_r h^2)^{-1}$ and is a measure of a 'mean free path' or cross section for Bragg reflection. Some values of ξ_g are given in Table 4.1 and it is clear that it is a function of the order of reflection and the atomic number of the scattering atom. These oscillations explain the thickness fringes and fringes at planar interfaces or defects often observed in transmission electron micrographs of crystals. These are discussed later in terms of the simpler but more approximate kinematical theory of diffraction.

One important effect that can only be explained by the dynamical theory is anomalous absorption which is essentially the limiting factor to electron transparency in crystalline materials. In reality it is an increase in inelastic scattering and a transfer of electrons from the Bragg reflections to the diffuse background of the diffraction pattern. This effects $\Psi^{(2)}$ more as it peaks at the atoms and hence it is scattered strongly by atomic excitation and phonon processes. $\Psi^{(1)}$ can penetrate crystals of the order of 10 ξ_g in thickness and hence in the correct orientation fairly thick crystals may be studied in transmission. Important contrast features, to be described below, are degraded in thick crystals essentially by anomalous absorption but this can be reduced by using electron beams of greater energy, as discussed in Chapter 6. It is clear that mass thickness concepts of transparency do not apply to crystals.

An approximate theory which finds many applications to the qualitative interpretation of contrast is the kinematical theory of electron diffraction. The major assumption here is that the intensity of the wave scattered from an atom or, indeed, a crystal, is small compared to the incident plane wave intensity. This condition usually implies that the crystal is thin, i.e. the number of scattering events is at a minimum and that the crystal is not at the Bragg position, see e.g. Figs. 2.9 and 2.10. This can be interpreted in terms of the well known Born approximation where the incident electron beam is at a potential V_i (typically 100 kV) which is large and much greater than the mean inner lattice potential V_0 of the crystal (typically 20 V) and the refractive index ($= ((V_i + V_0)/V_i)^{1/2}$) is a little greater than unity. Multiple scattering, i.e. dynamic events, is not allowed. A further approximation that has been used to simplify the mathematical aspects of the theory and which is experimentally convenient is the so-called 'two beam' approximation introduced above.

The amplitude of a plane wave scattered from an atom at a distance r_n from the arbitrary origin atom of the crystal (see e.g. Fig. 2.9) is given by

$$A_s = f \exp(-i\phi) \tag{4.5}$$

where ϕ is the phase difference given by Equ. 2.12 and thus

$$A_s = f \exp(-2\pi i (r_n \cdot g_{hkl})) \tag{4.6}$$

ELECTRON DIFFRACTION

Table 4.1 A list of some approximate values of extinction distance (ξ_g) illustrating its dependence on (a) atomic number Z and reflection (through the structure factor F_g) and (b) accelerating voltage (expressed here as ξ_{kV}). ξ_g is calculated from the constant C_2 of Equ. 4.10 and values of $f(s)$ given by Doyle and Turner (1968) and is expressed in nm.

Reflection			110	111	200	211
Specimen	Z	Structure				
Al	13	fcc		56	68	
Ag	47	fcc		24	27	
Au	79	fcc		18	20	
Fe	26	bcc	28		40	50
Si	14	diamond cubic		60		
			$10\bar{1}0$	$11\bar{2}0$	$20\bar{2}0$	
Mg	12	hcp	150	140	335	
Zr	40	hcp	60	50	115	

(a)

Specimen	Reflection	ξ_{50}	ξ_{100}	ξ_{200}	ξ_{1000}
Al	111	41	56	70	95
Au	111	12	18	24	28
Fe	110	20	27	41	46
Zr	$10\bar{1}0$	45	60	90	102

(b)

Here f is the atomic scattering factor (Chapter 2) a measure of the scattering cross section of an atom, and g is the reciprocal lattice vector corresponding to the crystal planes which give rise to the observed reflection. The summation of the individual amplitudes scattered into a particular reflection by the atoms in a unit cell $F_g = \Sigma_n A_s$ is called the structure factor and is of fundamental importance in the analysis of diffraction patterns. The arrangements of atoms within

74 CONVENTIONAL TRANSMISSION ELECTRON MICROSCOPY

the unit cell of a crystal structure can be such that certain characteristic reflections are absent. These 'fingerprints' can identify the crystal system, i.e. in the face centred cubic system h, k and l must be all odd or all even (a demonstration of this is given in the Appendix).

The amplitude scattered (or diffracted) from a thin crystal at a position some distance from the crystal can be conveniently calculated in relation to Fig. 4.8. An electron wave is incident normally on a crystal and is scattered in transmission. To avoid inconsistencies in the calculation one must assume, as indicated above, that a finite intensity can be scattered if the crystal is oriented at an angle slightly different from the Bragg angle. This is observed experimentally and is best considered in terms of the reciprocal lattice. Fig. 4.8 illustrates the scattering of a wave K_d at an angle $(2\theta + d\theta)$ where θ is the Bragg angle. For this to occur the reciprocal lattice point at P must intersect the reflecting sphere and this is possible if the point is effectively extended into a line of length $2s$. This is consistent with the assumption of a thin crystal with incomplete destructive interference away from the Bragg position and the path

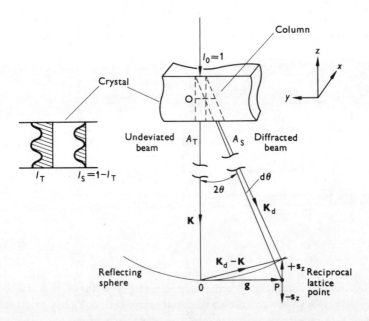

Figure 4.8 Scattering in transmission of a wave of unit amplitude incident on a thin crystal in the kinematic approximation. The concept of the 'column' and the oscillations of I_T and I_s through the crystal are illustrated schematically.

ELECTRON DIFFRACTION 75

difference is now $K_d - K = (g + s).r_n$, where r_n is the vector to each atom from an origin O, say, at the centre of the crystal.

The total scattered amplitude is therefore

$$A_s = \Sigma_n F_g \exp(-2\pi i [(g + s).r_n]) \qquad (4.7)$$

The product of g and r_n is an integer (i.e. g is a reciprocal lattice vector) and the effective phase factor is $2\pi s \cdot r_n$. If we refer the crystal, of volume V_c, to orthogonal axes, r and s can be written as $r = ux + vy + wz$ and $s = s_x + s_y + s_z$ and the total scattered amplitude for a crystal with $N = N_x N_y N_z$ atoms in the lattice is

$$A_s = C \sum_{u=0}^{N_x-1} \exp(-2\pi i (ux \cdot s_x)) \times \sum_{v=0}^{N_y-1} \exp(-2\pi i (vy \cdot s_y))$$

$$\times \sum_{w=0}^{N_z-1} \exp(-2\pi i (wz \cdot s_z)) \qquad (4.8)$$

These three summations can be obtained in like form which for the term in x is

$$\Sigma_u = \frac{1 - \exp(-2\pi i N_x x \cdot s_x)}{1 - \exp(-2\pi i x \cdot s_x)}$$

and

$$\sum_u \sum_u^* = \frac{\sin^2(\pi N_x x s_x)}{\sin^2(\pi x s_x)} \text{ for the intensity } A_s A_s^*.$$

The total scattered intensity is

$$I_s = C_1^2 \frac{\sin^2(\pi N_x s_x x)}{\sin^2(\pi s_x x)} \cdot \frac{\sin^2(\pi N_y s_y y)}{\sin^2(\pi s_y y)} \cdot \frac{\sin^2(\pi N_z s_z z)}{\sin^2(\pi s_z z)} \qquad (4.9)$$

(this result is obtained by Cochran (1973) for the case of X-rays). For a crystal thin in the z direction (which is parallel to the incident beam) and extended in the x and y directions such that s_x (and in the same way for s_y) $= (x N_x)^{-1} = (L_x)^{-1}$ is small (L is the dimension of the crystal), the first two terms of Equ. 4.9 approach constant delta functions and the scattered intensity varies only with s_z as

$$I_s = \frac{C_2^2 \sin^2(\pi N_z z s_z)}{(\pi s_z)^2}$$

76 CONVENTIONAL TRANSMISSION ELECTRON MICROSCOPY

Here s is assumed to be small so that $\sin \pi s_z z$ can be approximated to $\pi s_z z$ and writing $N_z z$ as the thickness of the crystal t the scattered intensity is

$$I_s = \frac{C_2^2 \sin^2(\pi t s_z)}{(\pi s_z)^2} \qquad (4.10)$$

The constant C_2 is equal to $\pi/\xi_g = hF_g/mvV_c$ when the intensity is normalized to take account of finite s_x and s_y and incident unit amplitude. ξ_g is known as the extinction distance for a particular reflection and is twice the distance in the crystal at which the diffracted beam reaches a maximum at $s = 0$, the Bragg position.

With the sum of the scattered amplitude and undeviated amplitude, A_T, restricted to unity, the maximum value of I_s is $(\pi t/\xi_g)^2$. The principal limitation of the kinematical theory is indicated here because for a crystal of thickness greater than ξ_g/π, $A_s > 1$ for small values of s. Hence for crystals thicker than an extinction distance (~ 10 nm) the kinematical theory is not really valid near the Bragg position.

From Equ. 4.10 it is clear that I_s oscillates with s as shown in Fig. 4.9. The scattered intensity is peaked at $s = 0$ and the subsidiary maxima can usually be ignored. The width of the central maximum is $2/t$ and as t decreases the intensity broadens and the reciprocal lattice point effectively extends in a direction perpendicular to the thin crystal. This result is significant in that the reflecting sphere can often intersect many reciprocal lattice 'spikes' and the

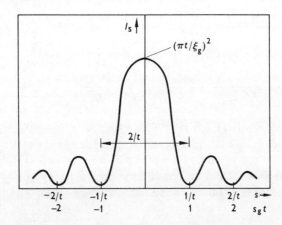

Figure 4.9 The oscillation of the scattered intensity in reciprocal space according to the kinematic theory.

ELECTRON DIFFRACTION 77

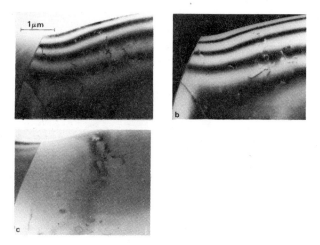

Figure 4.10 Thickness extinction fringes in a wedge shaped aluminium crystal, (a) the fringes in bright field and (b) in dark field by selecting the (002) reflection. On tilting the crystal away from the Bragg condition the contrast is lost (c).

resulting effect is the excitation of many simultaneous reflections. The distribution of intensity in reciprocal space thus depends on the shape transform of the scattering crystal or objects, as in optical diffraction.

A simple contrast feature explained by Eq. 4.10 is the appearance of fringes at a wedge shaped edge of a thin crystal. This is shown in Fig. 4.10 for an aluminium specimen both in bright field and dark field taken with the (002) reflection operating. If the diffracted and direct, undeviated beams sum to unit incident intensity i.e. $I_s + I_T = 1$ (assuming of course that $s \neq 0$) then both beams are at a minimum or maximum, Fig. 4.8, at successive depths in the crystal (I_s is at a minimum at $t \simeq \xi_g$, neglecting the slight dependence of ξ_g on s discussed later). Thus fringes of complementary intensity will be observed running parallel to the edge of the wedge in both bright and dark field. The separation of the fringes depends on s and they can be used as an aid to measure the thickness of the wedge at $s = 0$ because each white fringe in bright field corresponds to $t = n\xi_g$ ($n = 1, 2, \ldots$). The fringes are washed out in the thick parts of the crystal because of anomalous absorption.

The dependence of the scattered or diffracted intensity on s is illustrated by the appearance of contrast features, termed bend contours, in buckled crystals. Fig. 4.11 shows schematically the origin of the contours at $s = 0$. The planes responsible for the contour can be identified by dark field imaging of the

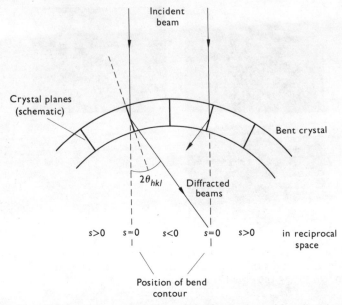

Figure 4.11 A schematic explanation of the formation of bend extinction contours in a bent crystal. Also shown is the variation of s in the vicinity of the contours.

pertinent reflection. The region of crystal for which $s = 0$ then appears as a light band joining points of equal bending running across the dark background. Bend contours imaged in bright field are evident in Fig. 4.12. Of course, more than one set of contours may exist depending on the number and identity of the reflections excited.

4.4 Contrast from an imperfect crystal

So far a perfect single crystal specimen has been considered and, apart from effects due to shape variations such as bending and changes in thickness, uniform intensity is expected in any particular image. A necessary requirement in microscopy is, of course, the observation of changes in intensity or the presence of contrast as this is the only way of detecting structural information. The presence of defects that are of an effective size greater than the resolving power of the transmission electron microscope can, in principle, be detected by changes in contrast that result from differences in electron scattering power between the defect and the surrounding perfect lattice of atoms. The simple theory outlined

CONTRAST FROM AN IMPERFECT CRYSTAL   79

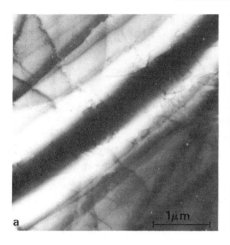

Figure 4.12 (a) A bend contour from (111) and ($\bar{1}\bar{1}\bar{1}$) reflections in stainless steel imaged in bright field. The image taken using the (111) reflection in dark field is shown in (b). Higher order contours are visible in (a) and increased transmission is seen on each side of the main contour (this is a dynamical effect for $s > 0$).

in the previous section enables this change of contrast to be calculated, at least in a qualitative sense.

4.4.1 Lattice defects

The contrast observed arises because of an extra phase factor introduced into Equ. 4.7 from a change in the lattice vector r_n when atoms are displaced by a vector $R(r_n)$, say, from their normal positions by the introduction of a defect into the crystal lattice. Equ. 4.7 can be written as

$$A_s = \frac{\pi}{\xi_g} \int_{\text{crystal thickness}} \exp(-2\pi i((g+s) \cdot (r_n + R))) \, dr$$

for small differences in phase. This relation may be modified by taking into account certain useful approximations. If the distortions are small, which is usually the case (i.e. $|R| \sim d$). and if s_x and s_y can be neglected for a thin sheet of crystal the phase factor $\phi = 2\pi(g+s) \cdot (r_n + R)$ approximates to $2\pi(g \cdot R + s \cdot r_n)$; as before $g \cdot r_n$ is an integer and $s \cdot R$ is small ($s = 1/t$ near the Bragg position and $d/t \sim 0.01$). The scattered amplitude is thus obtained by an integration through the thickness of the crystal along z;

$$A_s = \frac{\pi}{\xi_g} \int_{\text{crystal thickness}} \exp(-2\pi i(sz + \mathbf{g} \cdot \mathbf{R}(x,y,z))) \, dz. \tag{4.11}$$

The amplitude or intensity scattered from a defect is often calculated using an approximation known as the 'column' approximation. Because Bragg angles are small in electron diffraction the undeviated and diffracted waves A_T and A_s emerge from the bottom of a crystal almost parallel to each other, Fig. 4.8. If the deformed region of the crystal is divided into narrow columns then the displacement vector $\mathbf{R}$ can be assumed a function of z only, i.e. $\mathbf{R}$ is independent of the lateral coordinates. A_T and A_s then interact only with the strain field of the defect along a line essentially perpendicular to the crystal. Because θ_B is small the column can be very narrow $\theta_B t \sim 2nm$ and the approximation $\mathbf{R} = \mathbf{R}(z)$ is a reasonable one. For X-rays where the displacement of K_d at the bottom surface of the crystal is much larger note must be made of the strain field dependence on x and y, i.e. wide columns are required. The final intensity at the bottom of the crystal is summed with the variation in phase through the column (because of the change in $\mathbf{R}(z)$) in mind.

In practice the resultant integrations are often complicated and graphical methods such as the 'amplitude–phase' diagram are often resorted to (see any textbook on physical optics for the basic principles and Hirsch *et al.* (1965) for applications). It should be noted that if exact results are required the dynamical theory should be used in any case.

As examples consider a planar defect, such as a stacking fault, and a line defect i.e. a dislocation. Whilst it is not the intention here to discuss the structure of crystal defects in detail (for further reference see the bibliography) some consideration of the atomic arrangements at our two example defects is necessary. Fig. 4.13 illustrates the stacking fault schematically. The stacking sequence of atoms is interrupted, say from ABCABCABC ... in the face centred cubic (f.c.c.) crystal system to ABCABABCA ..., and the macroscopic result of this is equivalent to the displacement of one wedge of the thin crystal and part of the column by $\mathbf{R}$ with respect to the other part. The amplitude scattered by the column is

$$A_s = \frac{\pi}{\xi_g} \left[\int_0^{z_1} \exp(-2\pi i sz) dz + \int_{z_1}^{t} \exp(-2\pi i (sz + \mathbf{g} \cdot \mathbf{R})) dz \right] \tag{4.12}$$

The phase factor $2\pi \mathbf{g} \cdot \mathbf{R}$ is constant for a particular crystal and type of fault. For the case of an f.c.c. crystal with a stacking fault on a $\{111\}$ plane $\mathbf{R} = \pm a/3 \langle 111 \rangle$ (i.e. a distortion of magnitude $a/3$, a is the lattice parameter, directed along the $\langle 111 \rangle$ lattice direction) and $\phi = \pm 2\pi (h + k + l)/3 = 2n\pi/3$ where n is an integer. If $\phi = 2m\pi$ (m is an integer) then $\mathbf{g} \cdot \mathbf{R} = m$ and the crystal wedges are

CONTRAST FROM AN IMPERFECT CRYSTAL 81

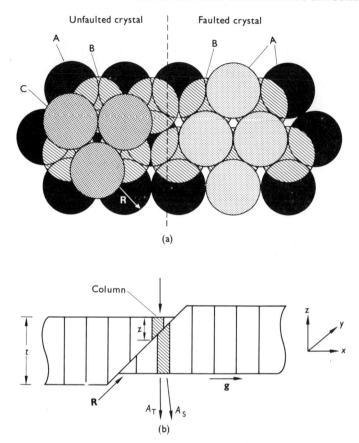

Figure 4.13 The stacking fault, (a) the atomic stacking sequence in unfaulted (ABC) and faulted (ABA) parts of an f.c.c. crystal. Displacement of the stippled atoms in the unfaulted (left hand) side of the crystal by R would produce the same fault as in the faulted (right hand) side of the figure. The crystal wedges and column are displaced by R at the fault (b).

back in register and no contrast is observed. Fig. 4.14 shows the fringe contrast observed at stacking faults in f.c.c. cobalt. Several faults are observed on different $\{111\}$ planes at different orientations to the electron beam. For $\phi = 0$ contrast is also not observed as $g \cdot R = 0$ and the phase factor is absent. This result is equivalent to saying that R is parallel to the Bragg reflecting plane (i.e. g and R are mutually perpendicular and g is always perpendicular to the real lattice plane). The absence of contrast at defects when this condition holds is a general feature,

82 CONVENTIONAL TRANSMISSION ELECTRON MICROSCOPY

Figure 4.14 Stacking faults in f.c.c. cobalt.

except for 'residual' contrast that occurs under certain special conditions. If the crystal is tilted, i.e. s varied, then the separation and contrast of the fringes changes.

Perhaps the lattice defect most often studied in transmission electron microscopy is the dislocation; the importance of this defect and its relation to other properties of materials such as mechanical and electrical properties is discussed in the references and bibliography quoted later. That dislocations should show contrast is evident from Fig. 4.15(a). Here the electron beam is incident at the Bragg angle on the displaced, strained planes of atoms to one side of an edge dislocation. It is clear that if the crystal is tilted the contrast will change and eventually disappear when the Bragg condition is not satisfied. The displacement vector R is a continuous function of z in the crystal in terms of b, the Burgers vector of the dislocation. In general the phase factor $\phi = 2\pi g \cdot R(b, x, y, z,)$ where the displacement R in the column approximation is determined at some point x, y, z in the column as shown in Fig. 4.15(b). The scattered amplitude in the general case is obtained as

$$A_s = \frac{\pi}{\xi_g} \int_0^t \exp(-2\pi i (sz + g \cdot R(b, x, y, z)))dz \qquad (4.13)$$

CONTRAST FROM AN IMPERFECT CRYSTAL   83

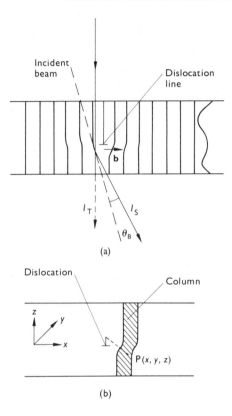

Figure 4.15 The dislocation, with (a) the schematic arrangement of planes at an edge dislocation and (b) the displacement of the column at the strain field of the dislocation.

For a whole dislocation $g \cdot b$ is an integer and for the special cases of $g \cdot b = 0$ no contrast occurs and again the displacement vector, or here the Burgers vector, lies in the reflecting planes. This condition is important as it is used as a means of determining b. By tilting the crystal with respect to the incident beam two different reflections in the two beam case are chosen for which the dislocation is out of contrast in sequential settings. The Burgers vector is given by the direction common to the two reflecting planes used, e.g. let a dislocation disappear for $(0\bar{2}2)$ and (111) reflections, the Burgers vector is then along $[\bar{2}11]$ i.e. $g \cdot b = 0\bar{2}2 . \bar{2}11 = 111 . \bar{2}11 = 0$. This situation is illustrated in Figs. 4.16(a) and (b) where the dislocation is in and out of contrast for different reflections.

84   CONVENTIONAL TRANSMISSION ELECTRON MICROSCOPY

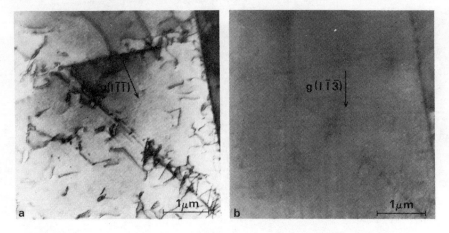

Figure 4.16   Images of dislocations in stainless steel. (a) The dislocations are in contrast for a ($1\bar{1}\bar{1}$) reflection and most are (b) out of contrast for a ($1\bar{1}3$) reflection. The diffraction vectors g obtained by comparison of the diffraction patterns with the micrographs are marked on the micrographs.

The simple diagram of Figure 4.15(a) suggests that the image of the dislocation should be to one side of the actual position of the defect. This is also clear from Equ. 4.13 where for different senses of x and/or y ϕ subtracts or adds to $2\pi sz$ and on each side of the dislocation the crystal lattice is turned towards or away from a reflecting position and the image is thus displaced.

This brief discussion can only give a few details of the contrast conditions in transmission electron microscopy in the study of line and planar defects. Information can be obtained about partial dislocations and the interaction of dislocations in deformed material. The crystallographic nature and origin of the important line defects known as dislocation loops can be determined. In the case of partial dislocations, fractional values of $g \cdot b$ can be obtained (i.e. b is not necessarily a whole number of lattice spacings) and the invisibility criterion $g \cdot b = 0$ is not sufficient to determine b since loss of contrast can be obtained for other values of $g \cdot b$. For a discussion of this point and 'residual' contrast the reader is referred to more advanced treatments such as that given by Hirsch *et al.* (1965).

Dislocation loops can be formed during work hardening and deformation processes, in quenched materials and as a result of irradiation damage. They result from interaction mechanisms in long dislocation lines or from the condensation of point defects (for a detailed treatment see Henderson (1972)). An

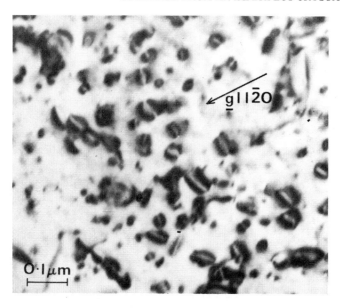

Figure 4.17 A micrograph of dislocation loops in neutron irradiated Zircalloy 2. The operating reflection (two beam condition) is $g = 11\bar{2}0$. (Courtesy of A. Riley and P. J. Grundy, 1972, *Phys. Stat. Sol. (a)*, **14**, 239.)

example of images of dislocation loops in a neutron irradiated zirconium alloy is given in Figure 4.17. The nature of these loops i.e. whether they are vacancy (formed from a condensation of vacancies) or interstitial (formed from interstitial atoms) can be determined by fairly straightforward contrast experiments in tilting the crystal to $\pm g$ positions. Transmission electron microscopy has, in recent years, played an important role in characterizing the nature of irradiation damage in materials used in nuclear reactor technology (for a review see Makin in Valdré (1971)).

Dislocation densities can be computed from electron micrographs as long as the crystal or foil thickness is known so that a true volume count can be obtained. Slip planes or twinning planes can be characterized and the nature of stacking faults can also be investigated by routine contrast analysis.

From all that has been said in this section it is clear that electron diffraction and diffraction conditions are responsible for the contrast features observed in the transmission electron microscopy of crystalline materials. Only a brief outline of the theory of contrast has been given but the message is clear; if a micrograph is taken without the diffraction conditions being known then the relationship between the contrast and its source can only be guessed at and is therefore open to error.

4.4.2 Precipitates and second phases

A major area of interest in materials science is the study of precipitation phenomena in the solid state and the structure of multiphase materials. A second phase often has a dramatic effect on the physical properties of a material and is often at the size level too small to allow an examination by optical techniques. This interest could comprise, for example, studies of precipitate identity, crystal structure, morphology, the kinetics of precipitation, precipitate sizes and dispersions and interfacial effects such as the problem of coherency and the characteristics of interfacial dislocations. This last defect can also occur in the important class of materials known loosely as composites where a composite structure of several phases is created by solid state reactions or mechanical methods. In the latter class, of course, the materials are often non-crystalline e.g. glass and carbon fibre material and polymers.

The fact that the matrix surrounding a second phase particle or precipitate is in the form of a thin section for transmission electron microscopy can cause complications in the interpretation of contrast and the computation of particle densities. This arises because the precipitate or second phase may be totally enclosed by the matrix or may be partially or almost completely uncovered and thus the diffraction conditions for crystalline materials can vary considerably. Precipitates can show contrast through several mechanisms, notably effects which can be associated with the particle itself such as orientation effects and structure factor contrast and also contrast from the interface and from the matrix itself. The contrast phenomena, which do not really have a universal terminology, are of course intimately linked with diffraction conditions and complicated effects can be introduced into the diffraction pattern of the material by the presence of a second phase.

Consider a small second phase particle larger than the resolving power of the microscope. If the particle is coherent with the matrix i.e. there is some continuity between the two crystals, as explained schematically in Fig. 4.18, then the matrix diffraction spots will be distorted; if the particle possesses a definite shape this distortion is predictable. As coherency is lost the two diffraction patterns separate i.e. diffraction from each phase is considered separately and distinct precipitate reflections are associated with the matrix diffraction spots. The occurrence of double diffraction, where the direct or diffracted beams in the matrix are diffracted again in the precipitate or vice-versa, complicates matters further.

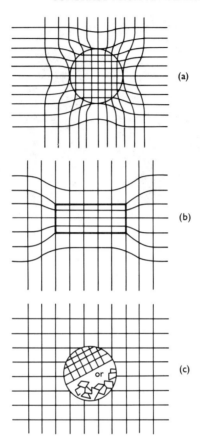

Figure 4.18 Schematic diagrams illustrating coherency effects between second phase particles and the matrix, (a) coherency; here the crystal lattices of the two phases are continuous and the spherical particle is surrounded by a spherically symmetric strain field, (b) partial coherency at a plate shaped particle and (c) an incoherent crystalline or amorphous precipitate (no continuity exists across the interface at the precipitate).

As for the actual observation of contrast, the type that shows features most analogous to those of dislocation contrast is that due to matrix strain fields i.e. the matrix is strained round a misfitting precipitate in the same sense that the matrix is strained near a dislocation. Fig. 4.19(a) shows strain contrast from spherically symmetric strain fields round a spherical precipitate of cobalt in copper. The displacement vector R is radial and contrast is observed everywhere

88 CONVENTIONAL TRANSMISSION ELECTRON MICROSCOPY

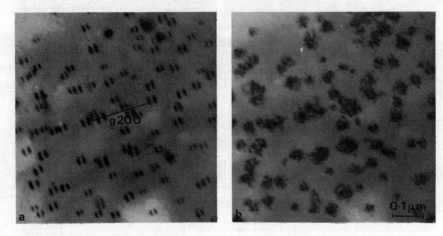

Figure 4.19 Micrographs showing the loss of coherency at the interface between cobalt precipitates and a copper matrix. (a) The strain contrast from the spherical cobalt precipitates. Note the absence of contrast perpendicular to g and (b) the loss of strain contrast after irradiation with 450 keV electrons (beam current = 5×10^4 A m^{-2}) for 5 minutes. (Courtesy of G. R. Woolhouse and M. Ipohorski, 1971, *Proc. Roy. Soc. Lond.* **A 324**, 415, and the Royal Society.)

except where $g \cdot R = 0$ and here a 'line of no contrast' results. Matrix contrast can also be observed from partially coherent plate shaped particles where, of course, the strain field is not spherical. In this case confusion can arise with dislocation loop contrast and the crystal should be tilted to try and introduce other forms of contrast such as orientation contrast. The misfit strain between a precipitate and the matrix can be relaxed by the generation of defects, vacancies or interstitials, at the interface. This process has important consequences as it can affect the mechanical properties of two-phase or multi-phase materials such as dispersion hardened alloys and directionally solidified eutectic alloys. In Fig. 4.19(b) the coherent cobalt precipitates of Fig. 4.19(a) have been irradiated with electrons and the resultant loss of coherency caused by the attraction of defects to the interface has changed the diffraction contrast completely. The contrast now originates in the precipitate itself and is probably of the structure factor type discussed below. It is clear from this example that a great deal of information can be obtained from the type of diffraction contrast observed and how important a clear understanding of the origin of any particular contrast is.

A characteristic parameter linking the crystal structure of a material and the diffraction conditions is the extinction distance ξ_g. A definition of ξ_g has already been given in terms of the diffracted intensity but it also varies from

CONTRAST FROM AN IMPERFECT CRYSTAL 89

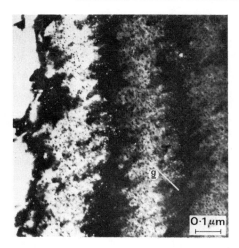

Figure 4.20 Small precipitates of TiO_2 in copper imaged by structure factor contrast. Note the displacement of the extinction contours, i.e. the reversal of contrast at the precipitates. (Courtesy of G. R. Woolhouse.)

material to material as it is a function of the unit cell size and the structure factor for a particular reflection F_g (see Equ. 4.10). These variations are a source of an important type of contrast. The presence of a small, coherent cluster of atoms or precipitate in a thin crystal effectively increases the thickness of the crystal by an amount related to the particle diameter or thickness D and the matrix and particle extinction distances ξ_g, ξ_g^p. This apparent change in thickness which, of course, is reflected in a change in position of any extinction contour is given by $t = \xi_g D(1/\xi_g^p - 1/\xi_g)$. The change in effective thickness produces a change in diffracted intensity in the column approximation and hence results in contrast at the particle. Fig. 4.20 shows this effect at particles of TiO_2 in an internally oxidized Cu–Ti alloy. The contrast is uniform over any particular precipitate and the displacement of the extinction condition to greater foil thickness at the precipitate is clear. The best contrast is obtained at the extinction contour, i.e. at $s \cong 0$, and this condition should be aimed for in the imaging of small second phase particles using this mechanism of contrast. This type of contrast is ideal for sizing and counting small precipitates. Good use of this contrast can be made in the case of materials which have been mechanically strengthened by the inclusion of small second phase particles either by internal oxidation (as in this example) or by precipitation hardening.

Often, misleading contrast effects are observed at precipitates and perhaps the most common of these is the appearance of fringes reminiscent of stacking fault fringes. Indeed, displacement fringe contrast is essentially the same as that of a stacking fault and arises from any remnants of strain at the matrix−particle interface. The phase shift ϕ is again $2\pi g \cdot R$ and unless the two parts of the matrix lattice above and below the precipitate are in register ϕ will have a finite value unless R is perpendicular to g. From the form of the fringes information on the shape of the particle, its habit plane and some idea of its identity (through the magnitude and sense of R) can sometimes be obtained. Other fringe contrast that is sometimes observed is that due to Moiré fringes. Their mechanism of formation is exactly that for their formation using optical gratings i.e. their relative separation and orientation are a result of the relative orientation and spacing of the two crystal lattice gratings of the matrix and the precipitate. This type of contrast can be used to study the arrangement of lattice planes at simple defects, such as dislocations, and the fundamental properties of epitaxy at crystalline interfaces.

Orientation contrast arises because of diffraction at a set of planes in the precipitate or in the surrounding matrix. The precipitate thus appears in uniform contrast and this contrast will depend on the angle of incidence of the electron beam. For this simple type of contrast the two crystal lattices must be unrelated, i.e. the precipitate, which is usually fairly large, must be incoherent or nearly so. The diffraction patterns from the two lattices are superimposed but independent and imaging of a precipitate reflection will cause reversal of contrast at the respective precipitates; in the case of structure factor contrast it is necessary to use a matrix reflection in dark field. This type of contrast is visually similar to that expected in an optical microscope image and is therefore relatively easily interpreted. An example of this type of contrast is given in Figure 6.12(a) where orientation contrast is seen at some amorphous silica precipitates in a nickel foil. The circular extinction fringes are thickness fringes in the matrix following the line of intersection of the spherical particle and the surface of the crystal.

In a text of this size it is not possible, and indeed it is not the aim, to discuss particular examples of any application in detail. There are so many publications and articles on the practical use of diffraction contrast in the investigation of materials problems that the consideration of one or two examples would be an injustice to many original and important pieces of work. The references cited in this chapter as well as some of the texts listed in the bibliography give a fuller discussion of specific applications in this field.

4.5 Specialized techniques in transmission electron microscopy

The aim of this section is to introduce several relatively new and often specialized techniques of transmission electron microscopy. The term specialized is meant to include techniques or modes of operation of the conventional microscope that have been recently developed and may in the future give new information as well as techniques that have a useful, but rather narrow, application to a particular phenomenon or class of materials. Recent applications of accepted and familiar modes to important new classes of materials will also be discussed. In a book of this size the choice of subjects is necessarily restricted and so the topics considered below should be viewed as one particular selection.

An important development in the study of defects in crystals at high resolution is the application of dark field techniques and the use of 'weak beam' conditions. Here the defect, i.e. the matrix strain field from a dislocation or precipitate, is observed in a dark field image using a low order reflection, i.e. one at small Bragg angle, with the crystal tilted away from the Bragg position ($s \gtrsim 0$). This situation corresponds to kinematical conditions and the defect is imaged at high resolution. The kinematical scattered amplitude at the defect is given by an equation of the form of 4.13. The integral is usually approximated by the stationary phase method and it can be shown that for a value of z where $(d^2\phi/dz^2) = 0$ the amplitude A_s reaches a maximum value. In the kinematical theory ξ_g becomes $1/s$ and a large s gives a small effective extinction distance and a narrow image width. Dark field conditions are used because the contrast is better than in bright field especially where the image almost disappears for $s \lesssim 0$. Application to stacking fault energy determination from the separation of partial dislocations bounding the fault is an obvious example for this technique and so is the study of stress fields at the interface of different phases.

As discussed in the previous section, atomic displacements exist at stacking faults and these displacements introduce a phase term into the expression for the scattered amplitude. The fringes observed in the images of stacking faults are often termed α-fringes. Other types of defect or deformation can introduce fringes into the electron image and the origin of the fringe pattern is often complicated. Antiphase boundaries (A.P.B.'s) in such systems as ordered Cu_3Au and domain boundaries in ferroelectric and some antiferromagnetic crystals are a source of such fringes. In the first case boundaries exist within a f.c.c. crystal between regions where the sublattices occupied by the Au atoms differ and in the second two examples boundaries occur between regions where the ordering of atomic or molecular electric dipoles and atomic magnetic moments is present but 'changes step' across the boundary. This ordering produces regions of crystal with differing symmetry properties whereas α-fringes originate from the relative displacement of identical parts of the crystal. An example of contrast at an A.P.B. boundary is given in Fig. 4.21. These fringe systems are dependent on their imaging conditions and often behave differently in bright and dark field settings. It is clear therefore that the observation of a fringe system or boundary

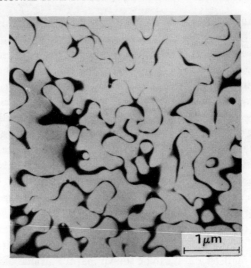

Figure 4.21 A dark field micrograph of antiphase boundaries in a Heusler alloy of nominal composition Cu_2MnAl. (Courtesy of A. J. Lapworth and J. P. Jakubovics, 1974, *Phil. Mag.* **29**, 253.)

in the image of a crystalline material does not imply the existence of a simple stacking fault, grain boundary or wedge. Further details of phase and domain boundary images are given by Gevers *et al.* (1964) and Amelinckx in Amelinckx *et al.* (1970).

So far in this chapter contrast has been obtained in the transmission electron microscope image by the use of a restricting aperture, the objective aperture. It is possible, however, as suggested in Chapter 2, to image related amplitude and phase information from the specimen. This information is only obtainable at high resolution and with the development of CTEM's with very high instrumental resolving power (0·2—0·5 nm) this mode of operation of the instrument is finding increasing application in materials science. The treatment given below is an extremely abbreviated one and extended discussions are given by e.g. Haine (1962) and Hawkes (1972).

A plane electron wave of unit amplitude travelling in the z direction is incident on the specimen. The incident disturbance at a point can be written as $\Psi = \exp(-2\pi ikz)$. The wave leaving the object on transmission is

$$\Psi_0 = T \exp(-2\pi ikz) \tag{4.14}$$

where T is the effect of the specimen on the wave through imposed amplitude and phase changes which are, of course, a function of position in the specimen.

The specimen function T is written as $T = a \exp i\phi$; a is almost unity as the amplitude is assumed to be hardly affected and can be written as $(1 - \beta)$ where β is small. The phase contribution ϕ is conveniently assumed to be small so that $\exp i\phi \simeq (1 + i\phi)$ (this approximation is reasonable in many cases for weak phase objects where the scattering is weak, or at low angles, over small distances). Neglecting products of β and ϕ, $T \simeq (1 - \beta + i\phi)$.

The disturbance in the image plane can be obtained from the object wave as $\Psi_i = A\Psi_0$ where A is often termed an 'aperture function' defining instrumental disturbances to the amplitude and phase of Ψ_0. If no objective aperture is present, in the sense that it does not cut out all but one beam from the image as it does in deficiency contrast, then A contains only a phase term as $A = \exp i\gamma$ and

$$\Psi_i = (1 - \beta + i\phi) \exp i\gamma \qquad (4.15)$$

Hence the image can be looked at as being formed from an amplitude 'specimen' β and a phase 'specimen' ϕ modified by an instrumental phase function in γ.

By analogy with optical diffraction from a grating the electron distribution in the Fraunhofer diffraction or scattering pattern is a Fourier transform of Ψ_0. The specimen can be assumed to consist of a superposition of periodic and non-periodic sinusoidal specimens with a distribution of detail spacings. The diffraction pattern, a frequency spectrum, consists of a set of discrete spatially positioned frequency maxima in the crystalline or periodic case or a continuous distribution in the non-periodic or amorphous case. The image is an inverse Fourier transform of the diffraction pattern and the function of the microscope can be expressed schematically as

$$\text{object} \xrightarrow{\text{F. trans.}} \text{diffraction pattern} \xrightarrow{\text{I.F. trans.}} \text{image}$$

The disturbance in the image is therefore a combination of amplitude and phase transfer and is a superposition of various spacings and frequencies. The microscope's response is characterized by an amplitude contrast transfer function, $-2 \cos \gamma$, for amplitude objects of strength β and by the phase contrast transfer function, $-2 \sin \gamma$, for phase objects. γ is composed of spherical aberration and defocussing contributions and for a wave scattered at an angle ψ to the optic axis the phase shift is

$$\gamma = \frac{2\pi}{\lambda} (\tfrac{1}{4} C_s \psi^4 + \tfrac{1}{2} \Delta f \psi^2).$$

In the conjugate image plane no contrast is observed if all transmitted electrons

94 CONVENTIONAL TRANSMISSION ELECTRON MICROSCOPY

are collected in the absence of aberration. The faithful transfer of information is thus dependent, for particular values of C_s and ψ, on the defocussing Δf. It can be shown that information from an amplitude object is only correctly transferred, over a wide range of detail, at certain non-zero values of Δf. At focus the amplitude contrast transfer function is zero for very fine object detail approaching atomic dimensions $\sim 0 \cdot 1 - 1$ nm. Also, the phase transfer function has plateaux corresponding to accurate transfer of phase information only over limited ranges of object detail for various values of Δf. These object spacings are of the order of $0 \cdot 1 - 1$ nm for typical Δf's of order 100 nm (for usual values of $C_s \sim 3$ mm and $\lambda \sim 4$ pm). The microscope therefore reproduces specific ranges of detail faithfully at a particular setting and others are distorted; however changing Δf adjusts the instrument to a particular spacing of choice. Details of these limitations are given in the references already quoted.

Figure 4.22 Phase contrast images of (111) lattice planes in a crystallized area of an amorphous germanium film. (Courtesy of S. R. Herd.)

Two important examples of phase contrast microscopy are in the imaging of structural detail approaching atomic dimensions and magnetic structure. The basic experimental requirements for the direct resolution of crystal lattice planes were introduced in Chapter 2 and the theory has been outlined above. Fig. 4.22 shows a micrograph of a vapour deposited amorphous germanium film partially crystallized in the electron beam. Lattice image fringes are observed in these regions and are the product of interference between the undeviated beam and the (111) reflection. Tilted illumination was used so that both the zero order (000) and (111) reflections were at equal angles of θ_B to the optic axis. A 40 μm objective aperture allowed the passage of both beams to the image. This micrograph was probably taken slightly out of focus as the fringes corresponding to (111) planes in germanium separated by about 0.33 nm are faithfully reproduced and are in good contrast. No fringe-like contrast is visible in the amorphous parts of the film. However there is some controversy in this area as some workers have detected crystal-like fringes in amorphous semiconductors and interpret the structure in terms of a microcrystalline model. This may be at odds with the results of electron diffraction but it must be remembered that electron diffraction gives structural information averaged over the diameter of the electron beam, typically 1 μm, and the thickness of the specimen and it is therefore possible to lose localized detail.

In a magnetic field B electrons are subject to a force F, the Lorentz force, given by $F = -e(v \wedge B)$ where v is the electron velocity and e its charge. If the object is a thin section of magnetic material in which the magnetic induction is B the resultant scattering or deflection angle resulting from the Lorentz force is $\psi_L = eBt/mv$ where t is the section thickness. For most magnetic materials ψ_L is of order 10^{-4} rad, a scattering angle much smaller than a typical Bragg angle of 10^{-2} rad and well within the objective aperture. Magnetic contrast is revealed by defocussing the objective lens and taking the distribution of electron intensity at a distance above or below the specimen as the object. This scheme is illustrated in Fig. 4.23 for typical magnetic domain configurations. The phase difference introduced by the specimen between two coherent waves travelling between conjugate points in the object and image and traversing the specimen at positions separated by Δx is given by $\phi = e\Delta F/\hbar$ where $\Delta F = \Delta B t \Delta x$ is the change in magnetic flux in the interval Δx. In reality the distribution in intensity below the specimen is a Fresnel interference pattern produced by interference between electron waves having a difference in phase ϕ resulting from diffraction at a change in flux ΔF. These phases are reproduced by the phase transfer function and for this reason the intensity variations in magnetic contrast are correctly described by wave optical treatments. However at small defocussing distances geometric approximations apply, as in optics, and the arrangement of the domain structure can be inferred. At large defocussing recognizable interference patterns occur and it is possible to use these patterns to investigate the detailed magnetic structure inside domain boundaries.

Apart from such fundamental applications Lorentz microscopy, as it has come to be known, enjoys applications in the elucidation of domain structures

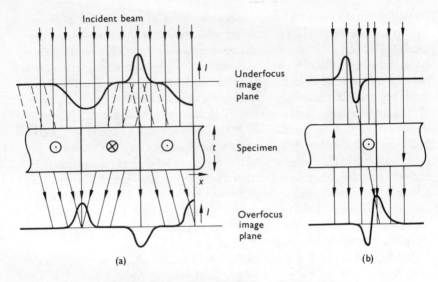

Figure 4.23 Schematic diagrams explaining the contrast from magnetic domains and domain boundaries and the reversal of contrast in the overfocussed and underfocussed images. Domains magnetized in-plane are represented in (a) and those magnetized normal to the plane in (b).

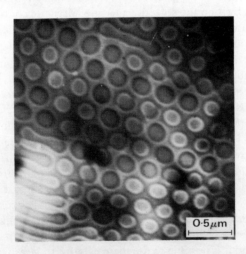

Figure 4.24 A Lorentz electron micrograph of cylindrical 'bubble' domains in a thin cobalt crystal. The contrast 'at' the domain boundaries is shown schematically in Fig. 4.23(b). (Courtesy of P. J. Grundy *et al*, 1971, *IEEE Trans Magn.*, Mag–7, 483).

SPECIALIZED TECHNIQUES IN TRANSMISSION ELECTRON MICROSCOPY

at high resolution in such technically important materials as those used in magnetic memory and logic areas in computer technology. One of these applications is in research into the magnetic properties of materials supporting a type of domain known as the magnetic 'bubble' domain. These cylindrical domains are mobile and can be used to represent information in a binary code, e.g. the presence of a bubble signifies unity and its absence zero. Lorentz microscopy afforded a detailed picture of the magnetization distribution in several types of boundary surrounding this kind of domain, Fig. 4.24. General and specialized references to Lorentz microscopy are Grundy and Tebble (1968) and Wade (1973).

5

The Scanning Electron Microscope and its Applications

5.1 Introduction

Some of the factors governing electron scattering and the consequent reasons for contrast in scanning electron microscopy have been outlined in Chapter 2. The aim of this chapter is to discuss these points in practice and to illustrate at the same time some of the possible applications of the various techniques. It must be realized that the field of application of scanning microscopy is wide indeed; the requirements for suitable specimens are much less stringent than for transmission microscopy and virtually anything that does not decompose or collapse in the beam or the vacuum of the instrument can be examined using emissive effects.

The SEM may utilize any of a number of different types of signal to produce an image from a specimen. In each case the microscope will be employed in a particular operating mode. Table 5.1 lists the operating modes most commonly used together with an indication of what information may be obtained. Some idea of the spatial resolution is also given. Subsequent sections will deal more fully with the contrast, resolution and applications of each of the listed modes presenting where appropriate, some practical details and experimental results. However the reader should be reminded that this book is not meant as a guide to instrumentation. Since the transmissive mode is now attaining a certain prominence separate discussion on this topic is deferred until Chapter 6.

There are two courses by which the applications of the SEM can be discussed. The more direct — and the one followed here in the first instance — is simply to follow Table 5.1 and take each mode in turn, demonstrating its capabilities. Unfortunately this approach tends to compartmentalize the applications of the device. The alternative (see Section 5.7) is to choose specific areas of interest e.g. surface or electrical properties and describe how the microscope can provide relevant information independent of any particular operating mode.

Table 5.1 Commonly used modes in the SEM

Mode	Type of signal collected	Contrast information	Spatial resolution
Reflective	reflected electrons	compositional crystallographic	100 nm
Emissive	secondary emitted electrons	topographic	10 nm
		voltage	100 nm
		magnetic and electric fields	1 μm
Luminescent	photons	compositional	100 nm
Conductive	specimen currents	induced conductivity	100 nm
Absorptive	absorbed specimen currents	topographic	1 μm
X-ray	X-ray photons	compositional	1 μm
Auger	Auger electrons	compositional	1 μm
Transmissive	transmitted electrons	crystallographic	1–10 nm

5.2 The reflective and emissive modes

These modes are discussed together because they are closely related as regards practical realization in the microscope. In the emissive mode the grid of the collector is maintained at a potential of + 250 V whereas the reflective mode uses a grid potential of about − 25 V, the negative bias preventing the collection of low energy secondary electrons.

5.2.1 Topographical and atomic number contrast

From Chapter 2 it will be clear that the characteristics of both reflected and secondary electrons are sensitive to variations in atomic number (hence composition) and topography. However the reflective mode is much more efficient in detecting atomic number contrast while the emissive mode is used when topographical information is required. When used in the emissive mode topographic contrast is so strong that it may dominate any other contrast mechanism. The difference in resolution between the two modes shown in Table 5.1 arises because the secondary and reflected electrons originate from different depths in the specimen. It is true to say that the majority of observations undertaken with the SEM exploit the topographic contrast provided by the emissive mode. If anything, therefore, this is the conventional application of scanning microscopy.

Before taking a micrograph in the emissive mode several experimental parameters should be checked in order to optimize contrast conditions. The most important choices are of final aperture size, working distance, beam voltage, specimen tilt and scan time. Many of these factors and their influence on instrumental performance were dealt with in Chapter 3. Under ideal conditions and with a suitable conducting specimen a resolution of 10 nm should be obtained. One further important practical point; if the sample is non-conducting it must be coated with a thin layer ($\sim$ 20 nm) of gold—palladium or other conducting material in order to provide a path to earth for any surface charge induced by the beam. Such charging would naturally impair resolution.

It goes without saying that any aspect of materials science that is concerned with sufaces is likely to benefit from the emissive mode of scanning microscopy. Particular examples in metallurgy are precipitate morphology and fractography. Fig. 5.1 is a scanning micrograph of a metallographic section of a cobalt alloy containing extended precipitates of silica which have been produced by internal oxidation. The silica appears in light contrast and the cobalt as a dark background. The resolution and depth of detail with which the filamentary precipitates are revealed provides ample testimony to the power of the technique. At this magnification the optical microscope would show nothing like the same quality, mainly because of the inferior depth of field. The contrast observed in Fig. 5.1 arises from two principal sources. First, the lightness of the precipitates compared with the darker cobalt background is mostly due to the difference in secondary electron yield of the two materials. Silica being an insulating oxide has much the higher yield. This is not to be confused with atomic number contrast which might add a small contribution if sufficient higher energy electrons reach the collector. The second cause of contrast is topographical, arising from the angular variations that the precipitates make with the incident beam. This micrograph illustrates to some extent another advantageous feature of the emissive mode. The collector has a large acceptance angle for secondary electrons which means that detail can be obtained from re-entrant areas such as cracks or holes which are not in direct line of sight of the collector.

If the positive potential on the collector grid is reduced the fraction of higher

THE REFLECTIVE AND EMISSIVE MODES 101

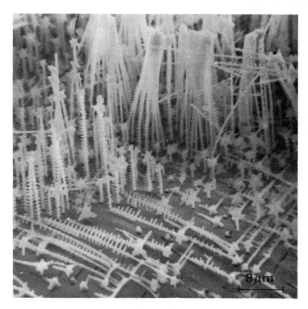

Figure 5.1 A cobalt alloy with a secondary phase of filamentary silica (emissive mode). (Courtesy of R. Barlow and P. J. Grundy, *J. Mater. Sci.*, 1970, 5, 1005).

energy electrons reaching the collector increases. Since these have linear rather than curved trajectories detail from re-entrant areas is progressively lost. Nevertheless in some cases the reflective mode can beneficially enhance topographical features and it is often worthwhile to investigate this possibility, especially in very flat surfaces. Figs. 5.2 (a), (b) are micrographs taken in the emissive and reflective modes respectively of a stainless steel sample which has been etched in hydrochloric acid. The triangular etch pits produced by the acid are clearly visible as is also a grain boundary at A. However in this particular instance no extra information is gained in the reflective mode but rather the loss of detail in the large central hole is exaggerated. Topographical information is lost from this region because the reflected electrons are not able to reach the collector. Despite the apparent limited efficacy of the reflective mode it is sensitive to changes in atomic number across the specimen surface. An example is shown in Fig. 5.3 of an iron–tantalum alloy containing a eutectoid phase.

Another area of study in which re-entrant surfaces are often found is that of fractography. Prior to the appearance of commercial scanning microscopes rough surfaces could only be examined at high resolution using replica techniques and

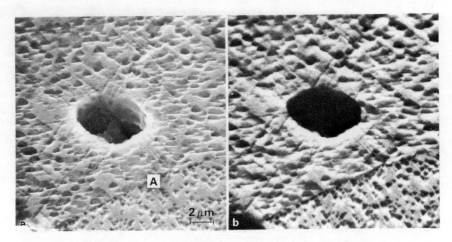

Figure 5.2 A stainless steel sample after etching in hydrochloric acid (a) emissive mode (b) reflective mode. Note the presence of etch pits, the grain boundary at A and the loss of detail in the central region of (b). (Courtesy of A. Riley.)

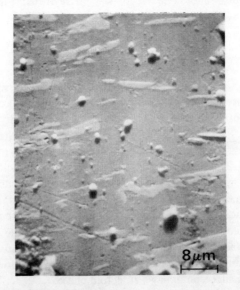

Figure 5.3 An iron–tantalum alloy containing a eutectoid phase (reflective mode). The extended regions of eutectoid phase appear in light contrast. The small bright globules are spurious deposits on the sample surface.

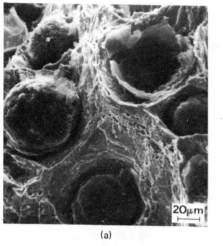 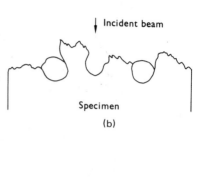

Figure 5.4 Ductile fracture in spheroidal graphite cast iron (a) emissive mode micrograph. (Courtesy of G. Jolley and S. R. Holdsworth.) (b) schematic representation of the sample surface presented to the incident probe.

transmission electron microscopy. Replication of very rough surfaces formed as the result of fracture is very difficult and sometimes impossible. Fig. 5.4 taken in the emissive mode is of spheroidal graphite cast iron in which ductile fracture has occurred. The ductile failure initiates at graphite nodules, several of which are seen in the micrograph as well as a site vacated in the fracture process. Between the nodules there is evidence of plastic deformation and tearing. Fig. 5.4 (b) gives a schematic indication of the specimen morphology presented to the incident beam. Micrographs of fractured surfaces can also reveal the presence or not of intergranular and transgranular cracking. If characteristic angles can be observed fracture planes may be identified.

The advantages of scanning microscopy in the study of fibres, textiles and polymers were recognized at an early stage. On account of their fragility and non-planar nature other methods of observation are much more difficult. Low accelerating voltages (~ 5 kV) are often used which means the optimum resolution cannot be obtained. The types of problem associated with textile research include the determination of the size distribution of constituent fibres and the study of the deleterious effects of washing, wearing and dyeing processes on fabrics. An added bonus of the SEM is that the large specimen chamber allows the possibility of dynamic experiments in stretching and fracture studies. Fig. 5.5 is an emissive mode micrograph of a sample of a 2-ply spun polyester sewing thread.

104   THE SCANNING ELECTRON MICROSCOPE AND ITS APPLICATIONS

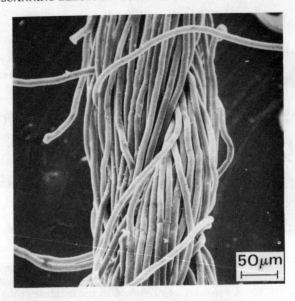

Figure 5.5   A 2-ply spun polyester sewing thread (emissive mode).

Unlike the CTEM, the bulk of observations made with the scanning microscope are qualitative in nature, the only measurements normally made being those of length. Nevertheless these need not necessarily be trivial and the correct depth and angular relationships of features found in scanning micrographs such as Figs. 5.2 and 5.4 can be obtained only by taking 'stereoscopic pairs'. A stereo pair consists of two micrographs taken before and after a small change in the angle of tilt — usually about 10° — or before and after a small rotation of the specimen stub. Superposition of the two images in a stereoviewer then gives a correct three-dimensional representation of the structure and allows the depths of 'valleys' and the heights of 'projections' to be ascertained. An account of the procedures involved in stereomicroscopy together with the geometric formulae required for the determination of dimensions has been given by Lane (in Hearle et al. (1972)). If, as is usually the case, the sample is tilted towards the collector there can be an appreciable difference in working distance and hence magnification between one edge and the other. This effect is most noticeable at the lower end of the magnification range. Stereomicroscopy also has applications in the CTEM, especially at high voltage, in determining the location of features through the thickness of the specimen.

Published results in many varied fields of materials science provide ample testimony to the versatility and efficacy of topographical contrast. In addition there is considerable activity in the life sciences. While recognizing therefore the eminent position of this somewhat routine type of scanning microscopy it is pertinent now to turn to more specialized techniques.

5.2.2 Electric and magnetic field contrast

Important contrast effects in the emissive mode arise from the presence of external electric or magnetic fields above the specimen surface. In both cases the trajectories of the secondary emitted electrons are altered by the external fields and clearly observed contrast may be obtained. Very little effect is observed in the reflective mode because the faster travelling back scattered electrons are not significantly deflected by the fields. Although field contrast is a comparatively specialized technique interesting results have been published in the study of magnetic domains and an example is shown in Fig. 5.6. The specimen here is of a ferrimagnetic material, yttrium orthoferrite, which contains a series of domains magnetized alternately into and out of the sample. The stray or magnetic leakage fields present above the specimen surface possess a spatial distribution related to the domain structure (Fig. 5.6 (b)). Above some domains therefore the emerging secondary electrons experience a force, the Lorentz force, which deflects them towards the collector while above others the electrons are deflected away from the collector. The contrast can be improved by replacing the grid over the collector with a plate containing a small aperture. Other factors which enhance contrast are a low accelerating voltage and a large probe size: the best spatial resolution obtainable is about 1 μm. Similar contrast effects to those described above may be produced by surface electric fields. Here again sensitivity can be improved by the use of restrictive apertures placed over the collector.

Another method recently developed permits the observation of domains in materials such as iron for which the external magnetic fields may be small. Back scattered electrons are collected which have suffered the Lorentz force actually inside the specimen. The contrast improves with accelerating voltage. Further experimental details of this technique together with an account of the present state of the art of scanning microscopy as regards magnetic contrast has been given by Fathers *et al.* (1973).

5.2.3 Voltage (potential) contrast

This is another contrast effect observable in the emissive mode and one which is particularly valuable in the study of semiconducting materials. If field contrast is understood as being due to the local electric fields which exist just above the

106 THE SCANNING ELECTRON MICROSCOPE AND ITS APPLICATIONS

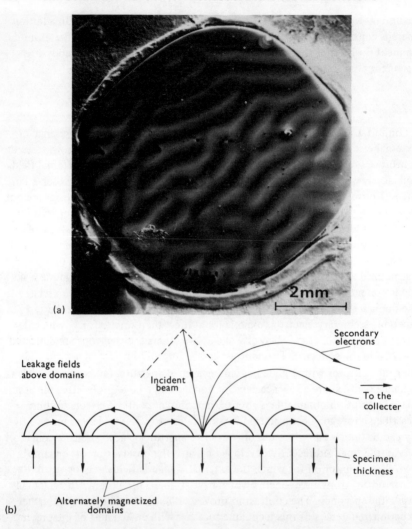

Figure 5.6 Magnetic domains in an orthoferrite sample, (a) emissive mode micrograph at 2 kV (Courtesy of P. Dunk). (b) a cross section through the specimen showing domains and the leakage fields above them.

specimen surface then voltage contrast is considered to arise because of potential gradients set up between the specimen and its surroundings e.g. the bottom of the final lens. These gradients, whose magnitudes depend upon the local specimen voltage, alter the energies and trajectories of the emerging secondary electrons. As a reasonable guide the lower the voltage at a particular specimen area the brighter will be the corresponding area in the CRT image. In an unmodified instrument voltage maps with a sensitivity of at least 1 V between adjacent regions can be obtained while still maintaining a reasonable spatial resolution. The best results seem to be obtained for untilted specimens and a low accelerating voltage. Various methods have been proposed for linearizing the signal i.e. making it proportional to the local specimen potential. These methods often involve the incorporation of additional potentials in the vicinity of the specimen and similar modifications to the electron collector assembly. Voltage contrast is not observed in the reflective mode because the collected electrons have too high an energy to be modulated by local specimen variations. As is usually the case in emissive mode operation the voltage contrast is likely to be overshadowed by topographical effects unless the specimen is flat.

The most widespread use of voltage contrast is found in the examination of semiconducting materials and devices e.g. the $p-n$ junction. If a $p-n$ junction with reverse bias is mounted in the microscope the p and n type material will be at different voltages and thus the junction region is revealed. The location of the junction in this simple way is often a prelude to the performance of more complicated experiments in the conductive mode. Apart from the simple $p-n$ diode, voltage contrast can be obtained in transistor and integrated microcircuits. Here the advantage of the SEM is that it will provide information about the device under actual operating conditions and allow its performance to be checked. Fig. 5.17 (b) shows a transistor with a reverse bias of 6·9 V applied between the base (at earth) and emitter: the bright contrast associated with the emitter region is clearly seen. Stroboscopic techniques have been developed to study the frequency dependent characteristics of these devices.

5.2.4 Electron channelling patterns (ECP's)

Electron channelling effects are one of the growth points of scanning microscopy because of the valuable crystallographic information they yield. Before turning to their applications we first consider their origin A more comprehensive account of many aspects of electron channelling effects has been given by Booker in Amelinckx *et al.* (1970).

Consider the probe striking a single crystal specimen at low magnification as shown in Fig. 5.7. Since the total angular excursion at low magnification is quite high, Bragg reflection may occur at various sets of lattice planes, e.g. AA', which are normal to the specimen surface. This results in the formation of bands of contrast on micrographs as illustrated in Fig. 5.8 which shows the effect in a single crystal of silicon. In terms of the deviation parameter s, see Section 4.3,

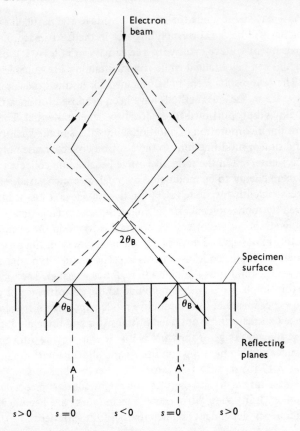

Figure 5.7 A simple experimental arrangement for the production of ECP's from a large single crystal. The dashed lines indicate the angular excursion of the scanning beam.

the exact Bragg condition corresponds to $s = 0$ while for the area between AA' $s < 0$ and for the area outside AA' $s > 0$. In this respect the bands bear a close relationship to bend extinction contours seen in the CTEM. The mechanism of contrast formation may be explained in a simple way as follows. In a simple approach to dynamical theory (Chapter 4) we may regard a diffracted beam as consisting of two types of superposed Bloch waves. The type (1) waves are peaked between the atom positions as they travel through the crystal whereas the type (2) waves tend to concentrate charge at the atom positions. This means that the back scattering cross section is much greater for type (2) than type (1) waves i.e. more reflected electrons (hence reduced penetration) arise as a result of interaction with the type (2) waves. It is a property of these waves that at the exact Bragg condition the two types are equally excited. However, just off the Bragg condition, for $s > 0$ Bloch wave (1) is preferentially excited while for $s < 0$ Bloch wave (2) is preferentially excited. Drawing the threads together we see that more reflected electrons will be produced in regions where $s < 0$ rather than $s > 0$ regions. This leads to the observation of Fig. 5.8 that the bands are bright with dark edges. A more rigorous theoretical analysis confirms this result. Although the reflective mode is often used to detect channelling patterns they can also be seen in the emissive and absorptive modes. Moreover a photographic film suitably placed in the vicinity of the specimen will record a 'channelling type' pattern if the scan generator is switched off. The term ECP (Electron Channelling Pattern) arises because the channelling effect essentially causes a dependence of the back scattered signal on the angle made by the incident beam to the lattice and so produces contrast.

ECP's can be obtained in a standard instrument although operating conditions are fairly stringent. Given a clean flat specimen the quality of the pattern is determined principally by the three quantities d_f, $\mathscr{I}$ and β (for the notation see Chapter 3). To be more explicit d_f determines the micrograph resolution, $\mathscr{I}$ the minimum contrast level and 2β the angular resolution. The first and second points have already been dealt with in Chapter 3. With regard to the third the best pattern resolution is obtained for $2\beta = 0$, i.e. when the electrons form a parallel incident beam. From Equ. 3.2 it can be seen that the three parameters are not independent and it is therefore not possible to obtain simultaneously good angular and micrograph resolution. The ECP of silicon shown in Fig. 5.8 (a) has an angular resolution of 2×10^{-4} rad but shows no surface topographic detail whatsoever. In order to observe the pattern well a contrast level of $C = 1\%$ is necessary. This requirement implies 10^6 electrons per picture point (Equ. 3.15) and currents of $\mathscr{I} = 100$ nA for video presentation. Some idea of the necessary operating conditions can be obtained by combining Equations 3.2, 3.15 and 3.16. They may be summarized as follows; low magnification, large probe current (> 10 nA), large probe diameter (> 1 μm) and low beam divergence ($< 10^{-2}$ rad). Improvement of ECP quality can be achieved by running the lenses at very reduced excitations. One simple method on a standard machine is to use the largest final aperture and to switch off the last lens.

Channelling patterns have very useful properties relating to the fact that they depend upon the crystallography of the specimen. The angular width of any band is $2\theta_B$ and so depends not only on crystal properties (i.e. 'd' spacings) but also upon the accelerating voltage which controls the electron wavelength. If V is known accurately then measurements of $2\theta_B$ yield the 'd' spacings and hence the

110   THE SCANNING ELECTRON MICROSCOPE AND ITS APPLICATIONS

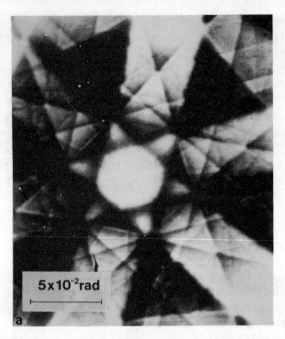

Figure 5.8 (a) An ECP obtained from a single crystal of silicon (Courtesy of E. M. Schulson *et al*, IITRI SEM Symposium, ed. O. Johari, 1969, 47).

lattice constants. Lateral movement of the specimen causes no change in the ECP but a tilt or rotation will produce change, a property shared with Kikuchi patterns seen in the CTEM (see Chapter 2 and the Appendix). Some authors refer to ECP's as pseudo-Kikuchi lines. By progressive tilting of the specimen in various directions an ECP map corresponding to the stereographic triangle can be constructed. This allows the crystallographic orientation of the specimen to be determined. If the pattern contains a low index pole it may be solved by inspection.

Apart from instrumental factors the quality of ECP's depends upon the perfection of the crystalline sample and this property has been made use of in various applications of the technique e.g. studies of in situ deformation and effects of radiation damage. Under certain circumstances grain contrast can be obtained from polished polycrystalline specimens, a situation which yields direct information about the size distribution of constituent crystallites. The microscope conditions required to obtain grain contrast are similar to those suitable for the observation of ECP's.

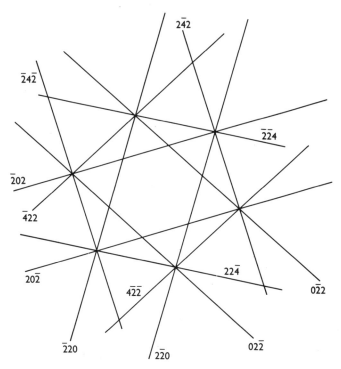

Figure 5.8 (b) indexing of pattern.

The great disadvantage of the operating conditions described thus far is that to achieve a large angle of scan, a low magnification must be used which in turn necessitates a large single crystal specimen. The potential value of channelling patterns from selected areas, analogous to those obtained in the CTEM will be readily appreciated. Two methods of producing ECP's from selected areas have been proposed and one is illustrated in Fig. 5.9. Instead of scanning the surface in the usual way the scan coil currents are modified so that the probe is rocked through a considerable angle about a point on the specimen. This motion produces an ECP from an area which may be only 5 μm in size. Obviously no topographical detail can be detected because the normal raster action is suspended. As the size of the selected area decreases the angular resolution also tends to decrease. This situation arises because to compensate for a small probe diameter, 2β must be increased (Equ. 3.2). A brighter electron source would improve matters. Selected area channelling patterns can be used to determine the crystallographic orientation of individual grains. Probably a more valuable

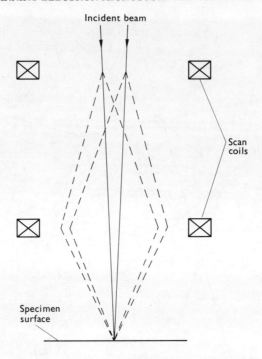

Figure 5.9 An experimental arrangement for obtaining selected area ECP's. The scan coil currents are modified to 'rock' the probe about a point on the specimen surface.

application rests on the fact that changes in the contrast, form and resolution of the patterns can serve to evaluate the density of lattice defects at particular specimen locations; for further details see Joy (1973).

Considerable effort has been put into the theoretical treatment of electron channelling effects. Naturally such an exercise requires the use of dynamical theory, but results in good agreement with experimental data have been obtained. As distinct from aspects arising with perfect crystalline specimens an intriguing question concerns the possibility of observing defects in the reflective mode image. Since the distribution of back scattered electrons is orientation dependent it seems feasible that dislocations near the surface could cause local changes in orientation and so become 'visible'. Theoretical analysis proves the soundness of this suggestion and predicts that dislocations and stacking faults should display oscillatory contrast which obeys the usual invisibility criteria (see Section 4.4.1). On the debit side, signal-to-noise considerations demand the use

X-RAY MICROANALYSIS AND AUGER SPECTROSCOPY

of sources and accelerating voltages which are brighter and higher respectively than those often available on standard machines today. With the introduction of field emission sources this situation will improve and back scattered images of crystallographic defects have already been observed. In agreement with theory these images are more easily obtained in thin films than bulk specimens.

5.3 Absorbed currents

If an electrical lead is connected between an illuminated specimen and earth a current will flow, the magnitude of which varies from one point to another. This specimen absorbed current is the difference between the primary current $\mathscr{I}_p$ and the sum of the secondary and reflected currents $\mathscr{I}_s + \mathscr{I}_r$. It is fairly small ($\sim 100$ pA) and therefore requires considerable amplification, a task for which commercial microscopes can easily be equipped. Since the specimen current is $\mathscr{I}_p - (\mathscr{I}_s + \mathscr{I}_r)$ its size for fixed $\mathscr{I}_p$ depends upon $\mathscr{I}_s$ and $\mathscr{I}_r$ which themselves depend upon the secondary and reflected emission coefficients. In other words the information gained from the absorptive mode is largely complementary to that provided by the reflective and emissive mode operation. However compositional variations are enhanced at the expense of surface topography insofar as effects due to specimen–collector geometry are eliminated. The resolution in this mode is somewhere in the region 20 nm–1 μm and is mostly limited by signal-to-noise effects. It is important not to confuse the absorptive mode with the more widely used conductive mode which exploits induced conductivity in semiconducting materials.

5.4 X-ray microanalysis and Auger spectroscopy

The production of X-rays by electron beams for analytic purposes is already well known in the electron probe microanalyser. In order to excite characteristic X-rays from an element it is necessary for the accelerating voltage to exceed, preferably by a factor of about three, the critical excitation potential, e.g. 17.5 kV for the Kα radiation of molybdenum. With an accelerating voltage of, say 30 kV, sufficiently intense K radiation can be excited in atoms up to $Z = 40$: for higher atomic numbers L and M radiations can be excited. The comparatively large voltages used mean that the sampled volume is of the order of 1 μm^3.

Once created the X-rays must be analysed. This can be done in two ways known as wavelength dispersive (WD) and energy dispersive (ED). The comparative merits of these systems can be mentioned but briefly; further details will be found in a paper by Jacobs (1974). In the WD method a crystal spectrometer is used. A narrow cone of emerging X-rays impinges upon a crystal which disperses them in such a way that only photons of selected wavelength (those fulfilling the Bragg law) reach a counter. The counter emits a voltage pulse whose height is proportional to the energy of each X-ray photon exciting it. A series of interchangeable crystals with different lattice parameters is provided to accommodate a large range of emitted X-ray wavelengths (0·1 — 10 nm). Good resolution (about 10 eV) for distinguishing between neighbouring spectral peaks is available in the WD system; it can also cope with a high generation rate of X-rays. However the geometrical arrangement of the components is critical. In the commonest form of energy dispersive system a lithium drifted silicon detector is sited close to the sample and receives from it the whole wavelength spectrum of X-rays. The detector discriminates between the various energies (and hence wavelengths) falling on it and the results are fed to a multichannel analyser which displays the number of quanta corresponding to any particular X-ray energy. The ED system can accept simultaneously wavelengths from many elements, is very rapid and requires comparatively low probe currents (<1 nA). The resolution of a solid state detector is about 150 eV and — unlike the WD method — cannot be used for X-rays with energies less than 1 keV. There is a tendency for reasons of speed and convenience to favour the use of ED techniques in scanning microscope applications.

Various types of output signal are possible in the X-ray mode in either ED or WD systems. An X-ray image can be thrown on to the video CRT by choosing a particular X-ray wavelength or energy. The location of the element possessing this characteristic wavelength will be revealed as bright contrast in the image. A modification allows similar information to show on a line scan. Alternatively the electron probe may be focussed on a spot to provide a point analysis. The whole range of X-rays can be collected and analysed to allow indentification of the region in question. Fig. 5.10 shows part of the spectrum obtained in such a way from a region of a transistor. Peaks from three separate elements, aluminium, gold and silicon are simultaneously displayed. As well as identification of elements determination of the concentration of an alloy component at a point is also possible. The height of the peak I_a for a given radiation from one of the constituents is first found; the sample is then replaced by a standard of the pure element and the new peak height I (a) noted. The ratio of these intensities gives an estimate of the weight percentage of the element in the alloy when corrections dependent on background, atomic number, absorption and fluorescence have been made. If computer facilities are available with the X-ray detection system a correction program for these factors may be incorporated.

The advantages of X-ray microanalysis techniques combined with scanning microscopy can be readily appreciated and wide application in the field of

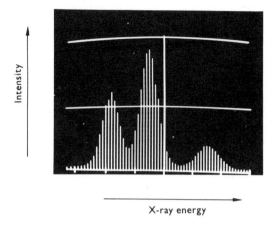

Figure 5.10 Part of the X-ray spectrum obtained from the surface of a transistor using an ED analysis system. The energy peaks from left to right are (i) aluminium Kα 1·5 KeV, (ii) silicon Kα 1·76 KeV, (iii) gold Mα 2·16 KeV. (Courtesy of L. J. Rabbitt.)

materials science has been made. For recent examples and more detail on the methods involved the reader is referred to the Bibliography.

There are difficulties associated with the detection of low atomic number elements for which the X-rays have low energies and long wavelengths (Fig. 2.20). Although the crystal spectrometer can cope, another technique used for the analysis of low atomic number materials ($Z < 11$) is that of Auger spectroscopy. Auger electrons have low energies (5–2000 eV) and are excited by accelerating voltages of about 2 kV. As a result the interaction volume is small and care must be taken to avoid surface contamination of the specimen. By the same token since penetration is low there is little spreading of the beam and good resolution is possible. The Auger electron spectrum may be displayed using a suitable detection system and thus allow compositional analysis to be done. As with X-rays a particular Auger electron energy can be chosen to display a distribution map on the video CRT.

5.5 Cathodoluminescence

Cathodoluminescence (CL) is a phenomenon which occurs in a great variety of materials ranging from biological specimens to semiconductors and minerals.

Before the discussion of applications it will be helpful to note a few general considerations that apply to its use in scanning microscopy.

Luminescent mechanisms take place throughout the interaction volume and hence the resolution is only about 100 nm. The emitted intensity increases with probe current and accelerating voltage but nevertheless is often still very low: this means that signal-to-noise problems can be severe. Enhancement of luminescent emission can be achieved by cooling the sample or using a brighter electron source. Other important practical points are (a) CL signals can only be obtained if the material under examination is transparent to the radiation being collected and total internal reflection does not constitute a barrier to escaping radiation, (b) the dwell time of the probe at any point on the sample must be greater than the relaxation time for the luminescent process otherwise blurring of the image will occur. Finally it should be noted that although light signals are collected the resolution obtainable is superior to that of the light microscope because there are no imaging lenses in the SEM below the specimen.

Contrast in micrographs may reveal variations in overall CL intensity or wavelength. An example of the latter is shown in Fig. 5.11 which is of a zinc selenide single crystal taken at a wavelength of 444·5 nm. The crystal has cubic and hexagonal phases together with polytype intergrowths. Recombination processes take place in all three structures which lead to cathodoluminescent radiation but of different spectral distribution. The use of a monochromator to select the 444·5 nm wavelength reveals the polytype intergrowths.

As mentioned in Chapter 2 cathodoluminescent intensity can be strongly sensitive to impurities and irregularities in a sample. These factors are of considerable benefit to the user of the SEM in the luminescent mode. By taking sequential micrographs in the emissive and luminescent modes the cathodoluminescent centres within a sample can be identified with features of the physical structure such as damage, defects or polytypic bands. By allowing some of the reflected primaries to reach the collector, a composite image can be obtained which gives luminescent as well as topographical information.

The link between impurities and luminescent centres has been widely utilized in the study of semiconducting materials. A good example is provided by the work of Shaw and Thornton (1968) who investigated Czochralski-grown crystals of laser quality GaAs heavily doped with 4×10^{24} Te atoms m^{-3}. CL micrographs of (100) oriented crystal slices revealed regularly spaced striations, Fig. 5.12, which betray the growth-revolution characteristic of this method of crystal fabrication. The CL contrast is due to variations in the concentration of dopant material induced by regular temperature fluctuations during the growth process; a diminution is found in the striated regions. Obviously this type of analysis can be extended to the study of dopant profiles in other materials.

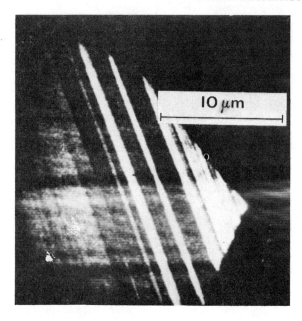

Figure 5.11 A luminescent mode micrograph using a discrete wavelength of 444·5 nm from a single crystal platelet of zinc selenide at 85 K. The bands of bright contrast are polytype intergrowths in the hexagonal phase, the luminescent intensity from which peaks at 444·5 nm. (Courtesy of P. M. Williams.)

The concentration of dopant is a factor of great relevance in the manufacture of such semiconductors as GaAs. For this reason the mechanisms responsible for cathodoluminescence in semiconductors and the SEM applications to these materials have been studied in some detail. A good account will be found in Thornton (1968). Very briefly the chain of events may be seen as follows. The incident beam striking the semiconductor produces electrons and holes which immediately begin to recombine. The CL efficiency depends upon the relative proportion of radiative to non-radiative combination. More specifically it depends upon the lifetimes of the two types of process. If the lifetime of the radiative processes (τ_r) is much less than that for the non-radiative processes (τ_{nr}) the efficiency will be high. Factors which cause a local variation in τ_r/τ_{nr} will contribute to contrast on CL micrographs. Such factors are impurities and defects. Of particular importance also is the surface of the semiconductor where non-radiative processes tend to predominate. Surface recombination is a function of beam voltage and minority carrier diffusion lengths (L_n and L_p).

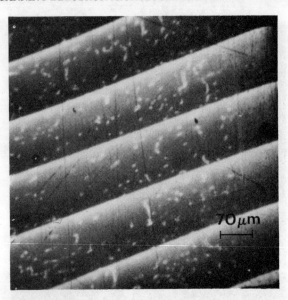

Figure 5.12 A luminescent mode micrograph of a single crystal of gallium arsenide, GaAs, doped with Te and grown by the Czochralski method. The regular striations are associated with a variation in the Te doping concentration. The bright spots with dark centres are edge dislocations. (Courtesy of D. A. Shaw.)

5.6 Induced conductivity

Beam induced conductivity has proved of great value in the investigation of semiconducting materials and is often used in conjunction with voltage contrast obtained in the emissive mode. The basis of the technique is the production by the beam of electron–hole pairs, i.e. the excitation of electrons from the valence to the conduction band which thereby leave holes. Under the influence of a bias field the holes and electrons drift or diffuse toward the appropiate electrode. The area across which the carriers move before recombining has an apparently higher conductivity which gives rise to current pulses in an external circuit under constant voltage. The process of charge separation with its resulting effect in an external circuit is often known as charge collection and the current produced as the charge collection current. Generally speaking those areas in which efficient charge collection occurs appear correspondingly bright in the image. Point to point contrast may be the result of variations in the generation rate of pairs (Equ. 2.21) or local changes in electric field, mobilities and minority carrier lifetimes. The resolution of this mode is mainly determined by the penetration and spreading of the incident beam as well as signal-to-noise considerations. Resolutions of 100 nm are possible. Apart from current impulses the diffusion of beam induced carriers can give rise to local changes in voltage at discontinuities and defects.

The experimental arrangement for exploiting beam induced conductivity is simple, provided a good amplifier is available (Fig. 5.13). The charge collection currents flow through a load resistance and the resulting variations in voltage are then amplified. The set-up of Fig. 5.13 can be represented by an equivalent circuit in which the $p-n$ junction is replaced by a condenser and a leakage resistance (Thornton 1968). The increased conductivity at the junction is equivalent to the condenser charging which in turn leads to current flow in an external load. Naturally the technique is not confined to the study of $p-n$ junctions.

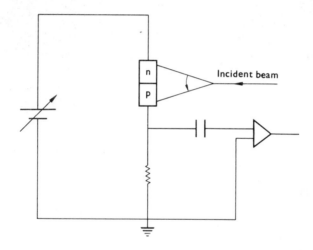

Figure 5.13 A simple experimental arrangement for the production of conductive mode contrast in a reverse biased p-n junction.

Two types of problem can be tackled with the conductive mode of scanning microscopy. In the first, the variation in charge collection current is used to probe structural features of the specimen. In the second, the behaviour of semiconductor devices such as diodes and field effect transistors can be examined under various working conditions. In this way the microscope is used as a diagnostic tool to detect possible faults and breakdowns in devices. Certain electrical measurements can also be made.

120   THE SCANNING ELECTRON MICROSCOPE AND ITS APPLICATIONS

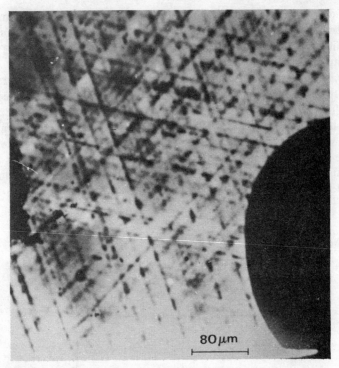

Figure 5.14 A diffused silicon diode taken in the conductive mode showing dark traces which are probably recombination centres associated with dislocations. The dark region on the right is a hole in the sample. (Courtesy of J. M. Titchmarsh.)

As an example of the investigation of structural information consider the micrograph in Fig. 5.14 which is of a diffused silicon diode with a $p-n$ junction some distance below but parallel to the surface. The irregular dark lines running along distinct crystallographic axes are localized recombination centres probably associated with dislocations. In the absence of these centres the image would be uniformly bright: in actuality the electrons and holes recombine at the dislocations before they have a chance to contribute significantly to the conductive signal. It is important to note that the microscope is being used to probe features which are below the surface of the specimen. Other structural features which can be revealed in the conductive mode are subgrain boundaries along which diffusion of p or n type material can occur. This local diffusion effectively changes the geometry of the junction which thus becomes apparent in the image. Apart from internal effects surface features, e.g. a film of contaminant, may affect the generation rate of electron—hole pairs and hence the charge collection signal. Normally the enhancement of the reflected electron fraction is accompanied by a reduction in the number of pairs since less energy for pair production is then available.

When studying the electrical behaviour of devices the interpretation of charge collection micrographs may not be easy. Most theoretical analysis has gone into

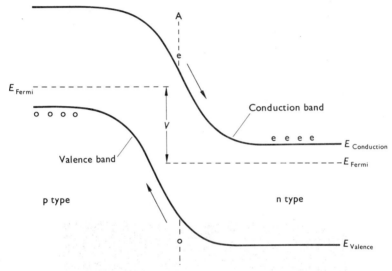

Figure 5.15 An energy band diagram of a reverse bias *p-n* junction showing the origin of conductive contrast. The incident probe creates an electron—hole pair at A which separates under the influence of the strong local electric field. V is the bias voltage. e represents electrons and o, holes.

the *p—n* junction (an energy level diagram for which is shown in Fig. 5.15) when operating under reverse bias. The conductive signal obtained from a junction depends upon its width and the minority carrier diffusion lengths of the materials on either side of the junction. Line scans taken across a 'vertical' junction in a diode at different reverse bias voltages are shown in Fig. 5.16. As the voltage is increased from zero the depletion region expands. However a signal is obtained not only from the junction region but from areas adjacent to the junction as minority carriers diffuse from there towards the junction field. Under certain conditions the slopes of the curves are $\exp(-x/L_n)$ and $\exp(-x/L_p)$ which allow the minority carrier diffusion lengths L_n and L_p to be determined experimentally. It is also possible in principle to obtain experimental

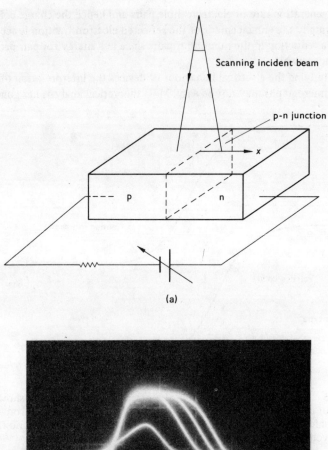

Figure 5.16 Line scans across a *p-n* junction in reverse bias, (a) experimental arrangement, (b) actual line scans for reverse bias of 0, 5, 10 and 15 V. The depletion region expands with increased bias. (Courtesy of L. J. Rabbitt.)

values for the minority carrier lifetimes. If the stationary probe is held at some little distance from the junction and then switched off the charge collection signal decays with time as $\exp(-t/\tau)$ where τ is the appropriate carrier lifetime. The measurement of τ is not quite as straightforward as that of L, requiring that the beam be switched off very quickly: this is best done by chopping the beam. Knowledge of both τ and L can lead to the determination of mobility μ through

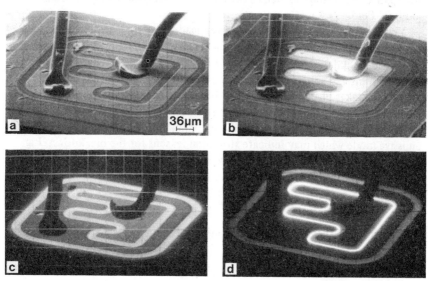

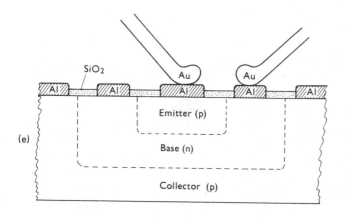

Figure 5.17 A p-n-p transistor with the base at earth; (a) emissive mode, zero bias; (b) emissive mode, voltage contrast, base–emitter bias = -6.9 V; (c) conductive mode, base–emitter bias = -3 V; (d) conductive mode, base–emitter bias = -6.9 V showing the onset of breakdown; (e) schematic representation of the transistor. (Courtesy of L. J. Rabbitt.)

Einstein's equation $L^2 = kT\mu\tau/e$. For details of other quantitative techniques the reader is referred to Thornton (1968).

Besides its use for measurements the SEM is widely used to investigate the properties of semiconductor devices. Fig. 5.17 is a series of micrographs of a $p-n-p$ transistor taken in various modes including (c) and (d) in conductive mode operation. The latter show the base—emitter junction contrast resulting from the applied bias voltage as well as the collector—base junction by virtue of its 'built in' field.

Very high fields applied to a device can lead to the passage of high currents which in turn may result in its breakdown. One common mode of breakdown is avalanche breakdown in which the electrons attain sufficient energy to ionize further atoms and so on. Those areas in a specimen where current multiplication of this sort takes place will be revealed because of a greatly enhanced charge collection signal. This property has been used to check devices for both bulk and localized breakdown effects especially in conjunction with examination in the emissive mode which may indicate structural reasons for the breakdown. Fig. 5.17 (d) taken at a reverse bias of 6.9 V indicates the onset of breakdown at the edge of the emitter—base interface. This behaviour is quite normal in devices of this type.

5.7 Conclusion. Techniques providing information in specific areas

It should now be apparent that as far as applications are concerned the SEM is an instrument endowed with considerable versatility. Despite this asset there is sometimes a tendency to rely too heavily on the conventional mode of operation. When tackling a particular problem therefore it is prudent to take an 'overview' and consider what extra information might be gained in other accessible modes. This section attempts to give a résumé of the techniques available for providing information on specific areas of interest. Broadly speaking the information obtained with scanning microscopy may be divided into four distinct types and these are compiled in Table 5.2 together with an indication of the appropriate modes of operation. This list must be seen as an adjunct to Table 5.1.

The emissive mode is normally used for the collection of surface topographic detail, contrast arising mainly from changes over the specimen surface of the angle of tilt to the incident beam. It is particularly effective for the study of rough surfaces but markedly less so for very smooth surfaces e.g. electropolished metallographic sections. In cases such as these the reflective mode is likely to prove more productive. A remarkable improvement in contrast is possible if the usual scintillator/light pipe collector is replaced by two semiconductor detectors situated above the specimen (Yeoman—Walker (1973)). By manipulation of the

CONCLUSION 125

Table 5.2   Types of information to be gained with the SEM operating in appropriate modes

Type of information	Modes
structural including topographic	emissive, reflective, conductive, luminescent, absorptive
chemical/compositional	X-ray, Auger, reflective, luminescent, absorptive
crystallographic	X-ray (Kossel), transmissive, reflective (ECP's)
electrical and magnetic	emissive, reflective, conductive

magnitude of the reflective signals from the detectors the structure of the smooth sample surface can be more thoroughly investigated. The structural information that may be obtained with scanning microscopy need not be confined to the surface only. An example of this assertion has been cited earlier; the probable observation of dislocations in silicon using conductive contrast. Cathodoluminescent signals are also structure sensitive and defects may be visible below the specimen surface.

Another principal advantage of the SEM is that it can be used to gain information about the composition of samples. The simplest technique is to collect back scattered electrons the yield of which is fairly sensitive to atomic number. Even in the emissive mode contrast between different types of materials e.g. insulators and metals can easily be seen (Fig. 5.1). Insofar as the absorptive mode signal is dependent upon changes in secondary and reflected electron emission coefficients it may also give compositional and topographical data. In certain cases this mode can provide more complete information than reflective and emissive images. As illustrated earlier in this chapter the intensity and spectral components of cathodoluminescence both in the visible and infrared regions are affected by the presence of impurity phases. However the most comprehensive data about compositional variation are those provided by X-ray microanalysis. Apart from the basic detection system peripheral software may allow the storage, analysis and treatment of spectra. This is an area of instrumentation in which improvements are continually being reported.

Techniques which yield crystallographic information are also in a state of constant development. The use in this respect of back scattered electrons to form selected area channelling patterns is already well established and the actual imaging of defects is likely to become so as various aspects of microscope design progress. A specialized X-ray mode is also available which can be incorporated into a standard microscope with little difficulty. The X-rays created when the incident probe strikes the specimen diverge from a virtual point source and are diffracted by atomic planes set at the appropriate Bragg angles. As a result, diffracted cones are produced which intersect a film placed above the specimen in a series of conics called a Kossel pattern. This pattern contains information relating to the crystallographic properties of the sample. Articles dealing with the uses, analysis and experimental realization of crystallographic techniques will be found in a conference report, *Scanning electron microscopy: systems and applications* (1973).

As a result of recent developments the emissive and reflective modes render possible the direct observation of magnetic domain structures in a great range of materials. Moreover the method has certain advantages over more traditional techniques e.g. it can reveal the internal domain structure. Apart from the domain patterns themselves the probe may be used to investigate the field distributions from recording tapes and heads. As far as electrical properties are concerned the emissive mode signal is sensitive to and can distinguish between surface field and surface voltage. Variations in these quantities are therefore made visible in circuits and devices. The cathodoluminescent mode and particularly the conductive mode are used extensively in the investigation of semiconducting devices, in many cases yielding valuable quantitative results.

Before closing this chapter the subject of signal processing will be mentioned since it affects the quality of SEM images. The ready possibility of processing exists because the signals associated with most modes of operation are collected as a continuously varying function of time and can therefore be treated electronically before any image is recorded. In the first instance the aims of processing are to increase the signal-to-noise ratio, thus improving contrast generally, and to effect discrimination between contrast from different sources. Certain electronic aids for improving contrast are usually incorporated in a standard machine: other more specialized devices for processing noise are available commercially. The differentiation of contrast sources is also accomplished by electronic manipulation of the collected signals. Surface topography and atomic number contrast can be separated by combining the signals obtained from detectors placed above the specimen. Voltage contrast and atomic number/topographic contrast have been separated by treatment of 'beam chopped' signals. One important aspect of processing is electron energy analysis. This topic with its relevance to transmission and scanning transmission microscopy will be considered in Chapter 6.

6
Recent Developments in Electron Microscopy

6.1 Introduction

Conventional transmission and scanning electron microscopy have their main use in standard modes of operation, i.e. the amplitude contrast technique and the study of topography in the emissive mode. However new generations of electron microscopes and electron sources are being developed commercially and the main impetus for this is probably the success of the scanning mode of microscopy and the accompanying bonus of signal processing. The harnessing of X-ray analysis in the SEM has already been discussed and the extension of this facility to the transmission mode is an obvious step, assuming of course that a sufficient X-ray flux can be obtained. The electron microscope thus becomes something of an analytical instrument and this development can be carried further by incorporating energy analysis equipment to study the energy loss spectra of transmitted electrons.

The range of operation of an electron optical instrument can be extended in a direction other than those discussed above and that is the use of higher accelerating voltages and hence more energetic electron beams. High voltage microscopy is, so far, mostly confined to the conventional transmission instrument and as a result of the cost of such instruments they are limited to a few laboratories. The principal advantage of high energy beams to materials science is that of increased transparency. However, some part of this increase can be obtained in a standard CTEM where the accelerating voltage is increased to, say, 200 kV using conventional, single stage acceleration and no doubt greater transparency will be obtained in STEMs working at 100 kV.

In this chapter we first discuss high voltage transmission microscopy and then analytical microscopy in the conventional transmission mode. This is followed by consideration of the STEM, potentially a very important instrument, and finally electron energy analysis. These developments are by no means the only recent achievements in electron microscopy but in the opinion of the authors are the ones most significant as far as the future of the subject is concerned. Some recent and likely future trends will be mentioned in the conclusion.

6.2 Conventional transmission electron microscopy at high voltages

This section deals only with the CTEM at high voltages, the application of scanning techniques to high voltage electron probes being considered in the section on STEM in this chapter. In the last ten years or so high voltage CTEMs have been successfully designed and built by several laboratories in universities and research establishments and the leading microscope manufacturers have built commercial versions. The technology of these machines is now well founded and they are already giving results not available at conventional voltages. A high voltage electron microscope can be defined as one working at and above about 500 kV. For accelerating voltages up to 200 kV it is possible to use conventional electron guns with a single stage acceleration in a normal CTEM assembly. Most high voltage instruments in operation at the present time work at voltages up to 1 MV but there are exceptions and some are in use or are being designed to use electron beams with energy up to 3 MeV or even 10 MeV.

What is the motivation for going to high voltages? The most important reason is that increased penetration of the electrons should enable thicker specimens to be observed in transmission. Also, the decrease in electron wavelengths should lead to better resolution. The possibility of using thicker specimens is extremely important in some materials science applications where it is clear that many defect structures and dynamic processes in very thin sections are not typical of the bulk material. This disparity in behaviour is due in the main part to the proximity of the two surfaces of the thin foil.

One would expect more energetic electrons to cause greater radiation damage to the specimen and this is indeed the case for displacement damage but not for ionization damage. The latter result has raised great hopes for the application of HVEM to biological materials and it is the aim of many investigators to study live cells and hydrated tissues in the high voltage CTEM. The freedom to do this lies essentially in the ability to use 'environmental cells' where the specimen is kept in its normal environment whilst under observation. The form of the damage wrought by the electron beam in living organisms is not yet clear but no doubt a great deal of effort will be applied to this problem. Environmental cells have wide uses for materials science as discussed later.

6.2.1 Electron scattering at high energies

Increasing the energy of the incident electron beam reduces the associated wavelengths as shown in Table 1.1. The effect of this reduction is to increase the magnitude of the extinction distances given by $\xi_g = \pi V_c / \lambda F_g$ (Section 4.3) and to decrease the values of Bragg angle. Some idea of the relative changes in ξ_g

and θ_B can be calculated from the values of ξ_g and λ given in Tables 4.1 and 1.1. Consequences of these changes in ξ_g and θ_B are that the separation of diffraction spots is reduced at normal camera lengths and that more reflections are usually excited at higher energies. Since the radius of the Ewald sphere is greater for smaller λ the probability of intersection with reciprocal lattice points and the excitation of more reflections is increased. This is so because even though the extension of the reciprocal lattice point or the half-width of the scattered maximum near $s = 0$ (Fig. 4.9) and ξ_g are related by $s = t^{-1} = (\xi_g/\pi)^{-1}$ and an increase in ξ_g results in a decrease in s, the relative flattening of the Ewald sphere is of more effect. Hence multiple beam conditions are more probable at higher voltages and the two beam approximation is more difficult to realize. Another result of a change to higher voltages and a decrease in θ_B is that the column approximation (Chapter 4) is better justified as the undeviated and scattered waves leave the bottom surface of the crystal more nearly parallel.

Inelastic scattering is reduced at higher voltages and this decrease accounts for the increased thickness of specimen that can be accommodated in transmission and also for a decrease in chromatic aberration. For single crystals inelastic scattering has been treated as an anomalous absorption described in terms of the effect of an imaginary addition to the real lattice potential (i.e. $V(r) + i\,V'(r)$) on the propagation of the electrons. The 'absorption' can be accounted for in an equation of the form of Equ. 2.1 where σ_s is replaced by an absorption constant, μ. The simple two beam theory shows that $\mu \propto v^{-2}$ (v, the electron velocity) and for a constant value of I/I_0 a transparency thickness can be defined as $t \propto v^2$. Hence the thickness dependence on accelerating voltage should be parabolic in form as shown in Fig. 6.1. Multiple beam effects make the situation more complicated because of the detailed interactions of the various Bloch waves but the curves of Fig. 6.1 are a reasonable approximation. Experimental measurements, some of which are included in Fig. 6.1, are in fair agreement but it must be remembered that they are often subjective because they usually involve the judgement of contrast, defined often as a constant I/I_0, at a defect or fringe system. However, it is clear that the transparency thickness increases by a factor of between 3 and 5 between 100 kV and 1 MV.

Similar considerations have been given to amorphous and polycrystalline specimens where the equivalent of Equ. 2.1 is written as

$$I = I_0 \exp -\left(\frac{N}{A}(\sigma_e + \sigma_i)\rho t\right) \qquad (6.1)$$

where N, A and ρ are as before and σ_e and σ_i are elastic and inelastic cross sections. Allowance for both single and plural scattering in thick specimens ($t > 100$ nm) shows that σ_i decreases relative to σ_e at higher voltages. Use of a

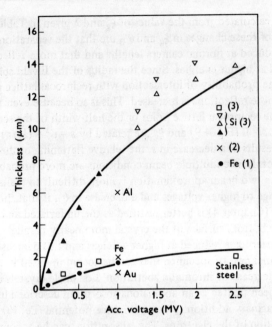

Figure 6.1 The expected form for the variation of transparency thickness with voltage in conventional transmission electron microscopy. The experimental points are reproduced by courtesy of (1) K. F. Hale and M. Henderson Brown (1970, *Micron*, 1, 434), (2) C. J. Humphreys *et al* (1971, *Phil. Mag.*, 23, 87) and (3) G. Thomas (1973, *J. Microscopy*, 97, 301).

transmission factor defined by $(\ln I_0/I)^{-1}$ gives fair agreement between the theory for plural scattering and experiment on carbon and gold films. However, it must be remembered that measurements of this kind depend on the size of objective aperture used, as this defines the total intensity collected.

As discussed in Section 2.3, other sources of energy loss apart from plasmon scattering are localized losses due to excitation or ionization of bound electrons and, at higher energies, atomic displacement effects. The last two losses can be bracketed together as sources of radiation damage. Ionization effects can be serious at 100 kV in non-conducting materials such as minerals, alkali halides, and polymers where breakdown of chemical bonds occurs. However, because of the reduced spatial rate of energy loss, improvement should occur at higher voltages. This is indeed found to be the case with many materials and very significant increases in specimen 'lifetimes' have been observed. The displacement of an atom from its normal site occurs when the energy transferred, E_t, from the incident electron is greater than the displacement energy E_d. E_d for the creation of an interstitial–vacancy pair in most metals lies in the range 20–30 eV. The transfer of energy E_T in an elastic collision at a scattering angle ψ is

$$E_T = \frac{4mME}{(m+M)^2} \sin^2 \frac{\psi}{2} \tag{6.2}$$

where m and M are the electronic and atomic masses. We can see that the fractional energy loss E_T/E in an elastic collision at small angles can be very small; assuming $M \sim 10^5 \times m$, $E_T/E \simeq 5 \times 10^{-5} (\psi/2)^2$ which for a typical Bragg angle $\theta_B \sim 10^{-2}$ rad gives $E_T/E \simeq 10^{-9}$. Equ. 6.2 can be approximated to $E_T = (2P^2 \sin^2 \psi/2)/M$ where P is the electron momentum. Writing P in terms of the relativistic kinetic energy and letting this energy be the minimum energy E_t to produce displacements then

$$E_d = \frac{2mE_t}{Mmc^2}(E_t + 2mc^2) \tag{6.3}$$

Approximate values of E_t (assuming $E_d = 25$ eV) are given in Table 6.1.

Table 6.1

Element	E_t (keV)
C	130
Fe	440
Ag	720
Au	1100

It is clear that at incident energies greater than 100 keV displacement damage can be a serious problem in the high voltage CTEM. However, the study of this effect for its own sake has proved of considerable importance as it offers a method of simulating the damage that occurs to materials in nuclear reactors. This topic is discussed further when some applications of HVEM are considered and has also been considered by Makin and Sharpe (1968) and Makin in Valdré (1971).

6.2.2 Instrumental points, aberrations and resolution

In high voltage electron optical instruments the dimensions of most components in the column are larger than at 100 kV. The working principles and construction of a high voltage CTEM are little different to those for a 100 kV instrument and the lenses are made to essentially the same design but are larger because 'thicker' lenses with greater magnetic fields are required to maintain the required optical properties with the more energetic electrons. The source of electrons is, usually, a conventional triode gun with sometimes a movable anode to obtain maximum working efficiency if the voltage range covered by the microscope is wide. The effective source of electrons from the gun is projected at the entrance point of the accelerator by either the first stage of the accelerator (acting as an electrostatic lens) or by an additional magnetic lens. The high voltage to the accelerator is provided by a Cockroft—Walton rectifier—capacitor type of generator and the acceleration of the electrons is effected in steps of about 50—100 kV in each step. Both generator and accelerator must be insulated against voltage breakdown and this can be ensured by air insulation in a specially designed room or by insulation using a pressurized gas such as freon or nitrogen. In principle, an unshielded high voltage electron microscope is a strong source of X-radiation and precautions to obtain adequate radiation protection for the operator are essential. This is provided by the fairly massive construction of the column and some extra shielding where the electron beam strikes apertures, the specimen and the viewing screen or photographic plate. Some instruments have facilities for remote operation from behind a radiation screen or for wheeling the operator's desk to a safe distance when large beam currents are being used or when the high voltage is being switched on and the instrument aligned. From Equ. 3.4 it is clear that a gun working at 1 MeV will give a greater beam brightness than at 1 keV but as the ionization cross section is reduced at higher voltages the images on the screen do not appear correspondingly brighter because the electron—optical conversion is less efficient. In common with some CTEMs at 100 kV high voltage microscopes often incorporate an extra intermediate or projector lens. One function of this extra lens is to increase the camera length of the instrument and so give adequate separation between diffraction spots when the Bragg angles are reduced at high voltage.

The aberrations of most importance in electron microscopy, namely spherical and chromatic aberrations have been discussed in Chapter 3. The aberrant disc at the object was given there as a combination of those due to spherical aberration and resolution in the absence of chromatic effects. The particular relationship chosen gave the best resolution as $r_{min} = 0\cdot 9\, C_s^{1/4} \lambda^{3/4}$. The decrease in wavelength at higher voltages should therefore give improved resolution, for a constant C_s, as shown in Fig. 6.2 due to Cosslett (1962). However, this improved resolution is not obtained in practice because the necessary increase in lens dimensions gives an increased C_s and technical problems with voltage stabilization near 1 MV introduce chromatic aberration in the electron beam.

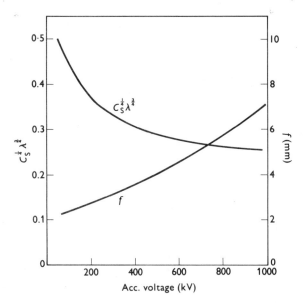

Figure 6.2 The dependence of resolution, defined as $C_S^{1/4}\lambda^{3/4}$, and focal length f for an objective lens on accelerating voltage. (Courtesy of V. E. Cosslett, 1962, *J. Roy. Microscop. Soc.* **81**, 1.)

The decrease in inelastic scattering at high voltages is reflected in a decrease in specimen induced chromatic aberration. The inelastic spread in wavelength, $\Delta\lambda$, and the equivalent spread in focal length of the lens gives a contribution to resolution which is proportional to $\Delta E/E$. For a disordered material the spread in energy ΔE is essentially determined by $\Delta E \propto NZt/mv^2$ where the symbols have their usual meaning. The loss ΔE and the spoiling of resolution by specimen induced chromatic aberration decreases by an order of magnitude between 100 keV and 1 MeV.

6.2.3 Areas of application

High voltage electron microscopy has been applied to many problems in materials science but it is clear that certain major interests have crystallized. These are (1) the utilization of thick specimens for studies of defects and 'difficult' materials and for structure-related, in situ dynamic experiments (2) experimentation with environmental cells to reproduce for example surface chemical reactions with the intention of studying the kinetics and products of the reaction in microscopic detail and (3) to take advantage of features peculiar to the HVEM such as the energetic electron beam for radiation damage studies. The terminology 'difficult specimens' applies to materials from which it is difficult to make good specimens suitable for 100 kV microscopy. As examples we may choose investigations typical of one or two of these categories.

Fig. 6.3 shows an example of a study of surface oxidation on α–titanium. The micrographs were taken at 800 kV with an atmosphere of 20 torr of air surrounding the titanium foil in the gas reaction cell. Fig. 6.3 (a) shows a grain boundary triple point junction at room temperature. Some hydride precipitates, introduced during electropolishing in the preparation of the foil, can be seen 'attached' to the grain boundaries. These precipitates dissolved during heating to the reaction temperature of about 600°C. Fig. 6.3 (b) shows the same area, at temperature, after 3 minutes. The stresses introduced by the oxide growth have produced extensive plastic deformation of the titanium which leaves the dislocations and slip traces visible in the left hand grain. After a total oxidation time of 20 minutes the metal is covered with a high density of 'islands' of preferentially thickened oxide, Fig. 6.3 (c). This non-uniform growth results in the production of a highly porous oxide as shown in the completely oxidized specimen.

It is suggested by this series of micrographs and other results that very valuable information can be obtained from the use of environmental cells in high voltage electron microscopy. Corrosion and gas reaction studies can be made at what are reasonable partial pressures of the reacting gases and microscopic detail of the chemical reactions can be obtained. A wide application to corrosion metallurgy, catalysis studies and surface physics and chemistry of this facility is obvious as evident from the reviews by Flower (1973) and Goringe (1973).

Soon after the development of the first research nuclear piles and reactors it became clear that an important requirement for the design of successful and economic power reactors was a knowledge of the radiation damage sustained by the materials making up the fabric of the reactor itself. Early experiments showed that neutron and fission fragment bombardment produced considerable dimensional changes in the fuel elements and containers from internal damage, and a consequent change in the physical properties of the materials in the irradiated regions resulted. Many studies of neutron damage in reactor materials have subsequently been made, see e.g. Fig. 4.17, but in experiments of this kind the specimen material for electron microscopy must be irradiated in the reactor for time periods of a month to many months to obtain a sufficient dose, i.e. number of irradiated particles per unit area, to simulate in-service damage. The current densities available in particle beams from ion sources and electron microscopes can be very large, e.g. for a CTEM it is of the order of 10^5 A m^{-2},

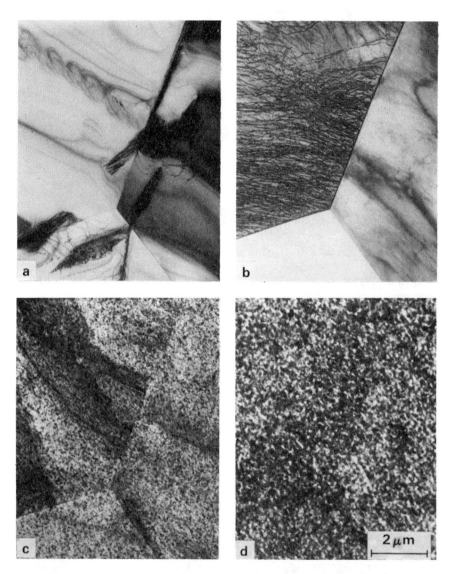

Figure 6.3 A series of micrographs illustrating an *in situ* study of surface oxidation on α-titanium. The foil is at (a) room temperature and (b) 600°C for 3 minutes, (c) 20 minutes and the foil is completely oxidized in (d). (Courtesy of H. M. Flower.)

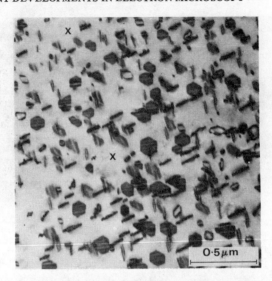

Figure 6.4 *In situ* irradiation damage in copper. The foil was irradiated with a beam current of $1 \cdot 2 \times 10^5$ A m^{-2} at 1 MeV. (Courtesy of M. J. Makin.)

and equivalent doses can be obtained in a time scale of the order of minutes. For example, a typical rate of irradiation in a nuclear reactor is 10^{18} fast neutrons m^{-2} s^{-1} and for a beam of the current density used to obtain Fig. 6.4 the rate of irradiation is 10^{24} electrons m^{-2} s^{-1}. Hence, providing the electrons have sufficient energy to cause displacement of atoms and condensation of defects is favoured in the specimen, the same results can be obtained in the high voltage CTEM in one minute as would be obtained from a reactor irradiation of 10 months. The damaging capacity of high energy electron beams can hence be used in the high voltage CTEM to give a detailed and controlled simulation of a nuclear reactor environment. Fig. 6.4 is a remarkable micrograph of irradiation damage in a copper crystal subjected to 1 MeV electrons at a beam current density of 1.2×10^5 A m^{-2}. The energy E_t, required to produce displacements in copper can be interpolated from Table 6.1 as being about 490 keV and so the incident electrons are energetic enough to cause displacement. The damage structures in the micrograph correspond to $0 \cdot 9$ displacements per atom, in other words 90% of the atoms suffer at least one displacement. The form of the damage is mainly faulted dislocation loops i.e. loops containing a stacking fault (see Henderson (1972)). These loops lie on $\{111\}$ planes and, as can be seen, the regular hexagonal shape with sides parallel to $\langle 110 \rangle$ of many of them suggests that the crystal is oriented very near to (111). Other loops are lying on three other {111} planes intersecting (111) and at some places, e.g. X, there is evidence for some defects, possibly loops, being nearly out of contrast for the diffraction conditions used. The ability to produce displacement damage in the high voltage CTEM is obviously a great boon to researchers investigating nuclear irradiation effects and presumably the development of higher voltage microscopes working at 3 MeV and possibly 10 MeV will produce striking results on materials such as uranium where the displacement threshold is about 1·2 MeV.

The high voltage electron microscope has already proved its usefulness to metallurgy, physics and materials science. It is usually in these sciences that new innovations and techniques are exploited quickly; presumably with the design of efficient environmental cells the life sciences will also benefit.

6.3 Analytical transmission electron microscopy

The incorporation of X-ray microanalysis facilities into the SEM has already been described in the previous chapter. The parallel advantages of X-ray analysis techniques combined with conventional transmission microscopy have also been recognized and various 'analytical electron microscopes' constructed to exploit them. It might be imagined that the simplest method of producing such a device is to place an X-ray detector somewhere near the specimen stage of a CTEM. However, the spatial resolution will not be very good unless the spot size at the specimen can be reduced considerably below that normally obtainable in a CTEM; this means modification of the lens system. Such a modification has been incorporated into the design of the analytical electron microscope EMMA–4 manufactured by AEI Scientific Apparatus Ltd and developed in conjunction with the Tube Investment Research Laboratories. This instrument probably has the highest spectral resolution of any commercial machine available. The normal imaging and diffraction capabilities associated with a high resolution 100 kV CTEM are available in EMMA–4 but its novel feature is an extra mini-lens situated between the second condenser lens and the specimen which when energized focusses a highly convergent beam on to the specimen. The size of the probe is less than $0·2$ μm with the probe current in the region of 10–20 nA. Since specimens suitable for transmission are so thin there is negligible spread of the probe, a factor which preserves the high resolution for analysis. The probe can be positioned anywhere on the specimen with the aid of deflection coils placed above the mini-lens. Two crystal spectrometers symmetrically disposed about the specimen enable the intensities of characteristic X-ray wavelengths to be monitored. The use of a mini-lens to produce a fine probe on the specimen is not the only method available. An alternative illumination technique adopted by some manufacturers is that of a 'condenser–objective' lens which in conjunction with the usual two stage condenser assembly gives a wide ranging variation in spot size suitable for both CTEM and analytical procedures.

The great advantage of an analytical microscope is that it allows a correlation of chemical composition with microscopic detail and diffraction data on a very fine scale. For this reason it is ideally suited to a whole range of biological and materials science problems where small concentrations of a minority element or precipitation are concerned. As an example we consider an investigation of the solute distribution in an iron-rich pyroxene taken from the work of Lorimer et al. (1973) using an EMMA–4. Pyroxenes are single chain silicates of Mg, Fe and Ca with the approximate chemical formula Ca (Mg, Fe) $Si_2 O_6$. Fig. 6.5 (a) is an electron micrograph of a calcium-rich pyroxene (augite) containing precipitate lamellae running diagonally across the specimen. In order to study the composition of the lamellae two independent traverses were made to determine the Ca/Si and Fe/Si X-ray intensity ratios. (The intensity of the Si radiation is expected to remain constant in the matrix and lamellae.) The traverses are revealed by contamination spots as seen in Fig. 6.5 (b); note the probe is made astigmatic to increase the resolution in one direction. The intensity ratios obtained are plotted in Fig. 6.5 (c) together with the position of the precipitates. Clearly the lamellae are calcium-rich and iron depleted compared with the augite host, an observation consistent with their being classed as another variety of pyroxene (pigeonite). Knowing the chemical composition of the augite host it was possible in this case on the basis of plausible assumptions to deduce the iron (21 wt%) and calcium (3.3 wt%) concentrations in the lamellae from the data of Fig. 6.5 (c). One important conclusion of this work is that unlike the situation with microanalysis in the SEM it is not necessary, because the specimens are so thin, to make corrections for X-ray absorption or fluorescence when carrying out quantitative procedures.

In further commercial development of a basic CTEM the condensing stage is adapted to produce a fine probe which is then scanned across the specimen, the image being displayed on a CRT. The instrument is now operating in a scanning transmission mode a subject dealt with in greater length in the next section. With the addition of X-ray facilities the machine becomes a scanning transmission analytical microscope. The spatial resolution achieved is again of the order of the probe size ($\sim$ 20 nm) since spreading of the beam inside the thin areas of the specimen will be minimal. The scanning action can be suspended and the static probe positioned at any desired point on the specimen. Hybrid microscopes with combined high resolution CTEM, STEM and X-ray microanalysis facilities are now available.

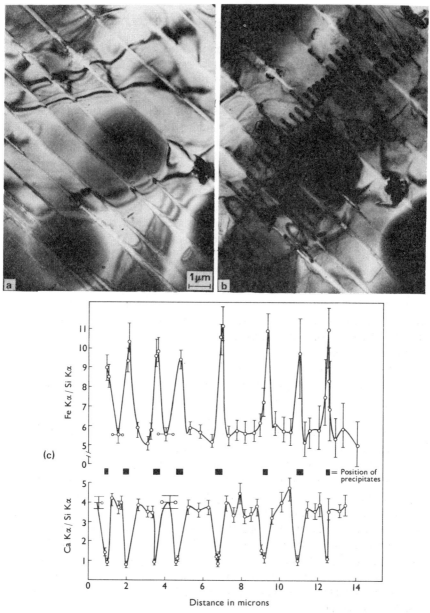

Figure 6.5 An analytical investigation of pyroxene, (a) Ion thinned pyroxene containing precipitate lamellae. (b) Same area as (a) showing contamination spots after two traces with EMMA-4. One trace measures Ca/Si ratios, the other Fe/Si ratios. (c) The Ca/Si and Fe/Si ratios obtained indicating that the lamellae are calcium-rich and iron-depleted. (Courtesy of G. W. Lorimer, N. A. Razik and G. Cliff, *J. Microscopy*, **9**, 153, 1973.)

6.4 The scanning transmission electron microscope (STEM)

The STEM is so named because it combines features of the two basic types of electron microscopes. Essentially it consists of a series of lenses which focusses a probe on to the specimen which is then scanned in the usual way. Unlike the normal SEM modes of operation however, the specimen is made sufficiently thin to allow the transmission of electrons. After transmission these are detected and the signals amplified and displayed. When STEM images were first obtained in the late 1960's a certain degree of surprise was aroused because they revealed lattice fringes and other contrast effects normally associated with the CTEM. Further examination shows that despite the dissimilar physical appearance of the two microscopes, as far as the electron optics is concerned, they are closely related and this suffices to explain the observed results. The basis of the explanation lies in a reciprocity theorem due to H. von Helmholtz which states that if a certain signal is detected at a point A when a source is placed at another point B, then the same signal in amplitude and phase will be detected at B if the source is placed at A. Two important qualifications of the theorem are (a) it does not apply to inelastic scattering processes unless the energy loss is small, in which case the intensities (not the amplitudes) are equal, and (b) it refers only to the interchange of source and collector; any apertures or similar means used to restrict ray paths must remain in position.

The application of Helmholtz's principle to the two microscopes is illustrated in Fig. 6.6. If the source of illumination and the condenser lens of the CTEM (Fig. 6.6 (a)) are replaced by a detector and the photographic plate exchanged for an electron source then an optical system corresponding to the STEM is obtained (see Fig. 6.6 (b)) provided the direction of travel of the electrons is reversed. The fact that different procedures are used for image detection is immaterial. According to the reciprocity theorem, since the ray paths are now identical similar images will be seen in both instruments if the specimen is inverted. As a result the contrast theory developed for conventional transmission microscopy should be of direct application to STEM images. Fig. 6.7 shows dislocations in silicon obtained with a STEM operating at 80 kV and with a probe size of 10 nm. Comparison of the two micrographs confirms that one set of dislocations in Fig. 6.7 (b) is out of contrast because the $g \cdot b = 0$ criterion is fulfilled for the (111) reflection: the close parallel with the CTEM is obvious (see Section 4.4.1).

In practical terms the instrumental resolution of the CTEM is largely controlled by the objective aperture angle 2α which, neglecting chromatic effects, is optimized for spherical aberration and diffraction (Section 3.6.1). Similarly for the STEM, the final lens aperture angle $2\alpha'$ is adjusted to produce the minimum

THE SCANNING TRANSMISSION ELECTRON MICROSCOPE (STEM)

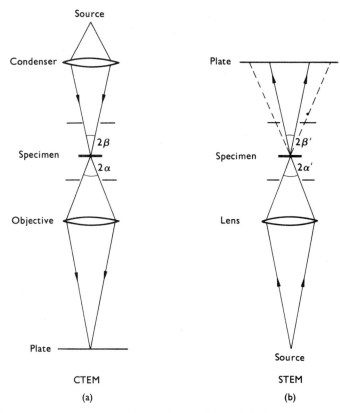

Figure 6.6 Ray diagrams for (a) CTEM and (b) STEM, illustrating the reciprocity principle.

probe size. The illumination angle in the CTEM cannot be increased indefinitely because of practical restrictions. On the other hand the equivalent angle $2\beta'$ in the STEM can be increased without loss of resolution. This is because the electrons transmitted through the specimen are not focussed by a lens and so chromatic aberration which arises from inelastic collisions and becomes increasingly severe with aperture angle is not relevant. It is therefore not necessary to operate the two instruments under conditions which are equivalent as regards reciprocity and a certain degree of latitude is available when choosing the collection angle $2\beta'$. For the case of amorphous material the maximum contrast is obtained in STEM images when $\alpha' = \beta'$, i.e. the illumination and collection angles are equal. The dependence on β' of the contrast of crystallographic features is somewhat more complex. In general the contrast falls with

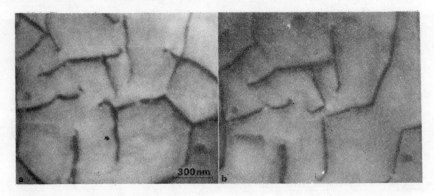

Figure 6.7 STEM images at 80 kV of a silicon specimen. Collection angle 10^{-2} rad. (a) taken in a (220) reflection, (b) taken in a (111) reflection. Note that one set of dislocations is now missing since $g \cdot b = 0$. (Courtesy of D. C. Joy, G. R. Booker, M. N. Thompson and W. Anderson.)

increasing β' — although the number of electrons contributing to the signal increases — but the rate varies with the type of feature under consideration.

Two approaches have been made to the practical construction of microscopes suitable for scanning transmission work. The first consists in modifying the types of standard machines currently available. Manufacturers of both the CTEM and SEM now provide stages which make this possible: in addition some prominent research groups have designed their own modifications. According to this philosophy the aim is to produce hybrid microscopes which serve several functions. The second course, adopted by other leading groups, is to design an instrument specifically for high resolution scanning transmission microscopy, perhaps particularly with biological applications in mind. Regardless of specific design any STEM to be used in materials investigation should possess certain basic facilities, notably (a) a good resolution micrograph mode in both dark and bright field and (b) a selected area diffraction mode. Below are indicated some methods by which these objectives can be achieved.

After passing through a crystalline specimen the electrons emerge along certain directions determined by the Bragg condition as illustrated in Fig. 6.8. (The scanning system and specimen position can be arranged so that the incident beam everywhere strikes the specimen approximately normally and keeps the diffraction pattern stationary at the plane AA'.) If the collection (or detector) aperture is placed to allow only the direct beam to enter the detector, a bright field STEM image is produced. The contrast arises because of point to point variations in the intensity of the direct beam as the incident probe scans over the

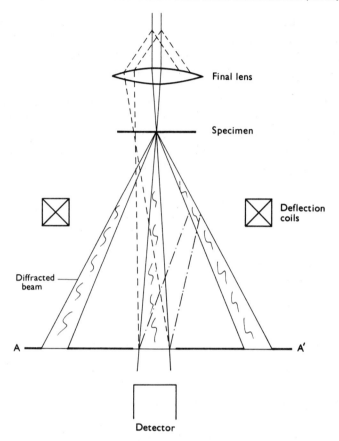

Figure 6.8 Bright and dark field imaging and diffraction modes in a STEM.

specimen. In principle if several detectors are available it is possible to display simultaneously on different cathode ray tubes both bright and dark field images. A more usual arrangement used to obtain dark field images is to deflect a given diffracted beam through the central aperture with a set of coils placed below the specimen. For non-crystalline samples, blocking off the unscattered electrons also produces a dark field image. As mentioned above the quality of the micrographs is influenced by the angles α' and β' while the magnification is controlled by the size of the raster on the specimen.

If a photographic plate is placed in the plane AA′ containing the collection aperture a diffraction pattern will be recorded. The diameter of the diffraction spots can be changed by altering the excitation of the final lens. With the scan coils in the final lens switched off and the stationary beam focussed on to the specimen the diffraction pattern from a selected area equal to that of the probe size should be obtained. Two techniques have also been proposed which enable scanning diffraction patterns to be displayed directly on a CRT. The first is a rocking technique similar to that described in connection with channelling patterns. Here the incident probe is held stationary on the specimen but rocked through an angle: the signal is provided by electrons which pass along the optic axis and through the centrally placed collection aperture. In the second method, which can also display Kikuchi patterns, the scan coils above the sample are disconnected and the diffraction pattern is scanned over the collection aperture with deflection coils situated below the specimen. Patterns from selected areas of the order of $1\mu m$ across may be obtained with either method.

Since scanning transmission microscopy is still at a comparatively early stage of development the resolution obtainable with modified instruments has probably not yet been optimized. A resolution of the order of 1 nm has been attained in a 100 kV CTEM modified to incorporate a scanning attachment. Useful references on STEM literature are listed in the Bibliography.

6.4.1 High resolution scanning transmission microscopy

Certain research groups and commercial manufacturers are pursuing the goal of designing a high resolution STEM, i.e. a device which will match the resolution of a good CTEM. Though details thus far published indicate that they have many features in common we shall concentrate on an instrument constructed by A. V. Crewe and his collaborators. This group pioneered high resolution scanning transmission microscopy and Crewe (1970) has written extensively on the subject. The main elements of the microscope are as follows (see Fig. 6.9).

(i) *A high brightness field emission source.* A high brightness is essential in order that sufficient current may be focussed into a very small probe size and yet provide an adequate signal-to-noise ratio. A field emission source is the only feasible means of achieving this aim. It consists of a single crystal tungsten wire whose tip gives off electrons in great numbers upon the application of a negative voltage. Unfortunately the environment of the tip must be maintained at ultra high vacuum (10^{-10} torr) but this disadvantage is outweighed by the fact that the effective source size is 10 nm or less in diameter. This leads to brightnesses in excess of 10^{12} $A\,m^{-2}\,sr^{-1}$.

(ii) *Electron gun.* This is designed to accelerate the comparatively low energy electrons emitted from the source to the high velocities required by the electrons to pass through the specimen. Keeping aberrations to a minimum the gun produces an image of the source about 10 nm in size a few centimetres below

THE SCANNING TRANSMISSION ELECTRON MICROSCOPE (STEM)

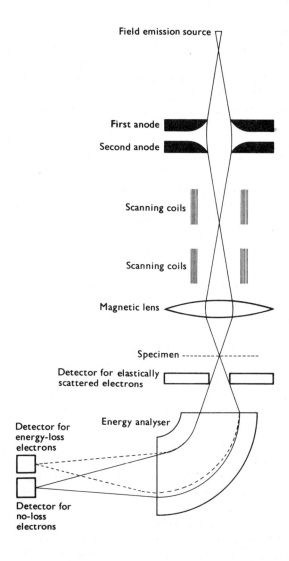

Figure 6.9 A high resolution STEM with electron separation facilities, built by A. V. Crewe and colleagues (from 'A high resolution scanning microscope' by A. V. Crewe, Copyright © 1971 by Scientific American, Inc. All rights reserved).

the second anode. This probe size would be ample for a conventional scanning microscope.

(iii) *Magnetic lens.* To attain a resolution comparable with the CTEM, the probe size must be further reduced and this is achieved using a short focal length lens (~ 1 mm) which demagnifies by a factor of 50–100. For such a demagnification factor the aberrations of the electron gun are negligibly small and the resolution is essentially determined by the aberrations of the magnetic lens. The situation is therefore equivalent to that in the CTEM where for a thin specimen the performance is dependent upon the quality of the objective lens. Thus the resolution limit (Equ. 3.13) applies in both cases. Crewe (1970) has reported a resolution of 0·5 nm in a carbon film at an operating voltage of 30 kV and resolved 0·34 nm lattice fringes in graphite.

(iv) *Energy analyser.* Clearly this component does not affect the probe size at the specimen but it does have an important bearing on the information obtainable. Three types of electrons leave the lower surface of the specimen, namely those that are (i) not scattered (ii) inelastically scattered (iii) elastically scattered through comparatively large angles. The latter are intercepted by the detector (Fig. 6.9) while the other two types pass through the detector aperture and enter the energy analyser where they are separated. It is possible in principle to display the three signals simultaneously. Valuable information can be obtained by analysing the scattered electrons as may be seen from the following consideration. The number of electrons scattered elastically or inelastically when the probe falls on an elementary region of the specimen depends upon the number of atoms within that region and their atomic number Z; (elastic scattering is more favoured at high Z, inelastic scattering at low Z). Let the probe fall on an element of a thin carbon film and the two scattered signals be combined in some way. If the addition of a single heavy atom to the element significantly changes the combined signal the extra atom will be rendered 'visible'. This technique has been used for the observation of single atoms of heavy elements. (Experiments along similar lines have also been done with the CTEM.) Besides this application there is the possibility of imaging unstained biological molecules and obtaining better resolved images of crystallographic defects.

As with the CTEM some attention has been given to increasing substantially the accelerating voltage in the STEM. The stimulus behind these developments is two-fold; an increase in electron penetration so that thicker samples may be examined, and the potential improvement in resolution. In order to gain the full benefit of the reduced wavelength however it is likely that some improvement will need to be made in electron gun and lens design. With increasingly thicker samples an extra complication arises especially in amorphous materials. Because of multiple scattering the resolution becomes progressively worse as the beam penetrates the specimen. This means that a feature on the upper surface will be imaged more sharply than one on the lower surface (the 'top–bottom' effect). Similar behaviour can be expected in crystalline films but in this case the magnitude will be orientation dependent.

Perhaps at this stage it is appropriate to consider why the development of a high resolution STEM should be of such great interest. The energy losses in a CTEM specimen are injurious to normal image formation; in a STEM on the other hand all the electrons, whether elastic or inelastic can be collected from a point and contribute to the signal. In other words, much more efficient use is made of all the transmitted electrons, a feature which has an important practical advantage because for the same accelerating voltage the STEM can cope with thicker specimens. (Of course lossy electrons can be harnessed to provide information in both types of microscope when combined with energy analysis facilities.) Another advantage of any scanning system is that the signals are collected electronically which offers great scope for signal processing and contrast expansion. All in all it is possible that the resolution and performance of the STEM will exceed that of the CTEM and for this reason is likely to become an important tool of electron microscopy in general.

6.5   Energy analysis and energy analysing microscopes

The physical processes involved in the scattering of electrons in solids have been discussed in Chapter 2 and the first section of this chapter. It is clear from these simple treatments that a study of the energy distribution of inelastic electrons in transmission can result in information relevant to chemical analysis. In this context energy loss can be a function of the atomic number of the scattering atoms and the electronic energy levels within these atoms. Information of physical and chemical importance such as the properties of the conduction electron plasma, revealed through inelastic plasmon scattering, is also available. The ability to obtain this information requires experimental methods of distinguishing or measuring the energy content of the transmitted electrons. Hence energy dispersive or selective facilities are required. Modular equipment of these types fitted to a CTEM or STEM convert the instrument to an energy analysing microscope. Here we consider the principles and scope of these techniques.

The electrons that emerge from the bottom surface of a specimen in transmission have a range of energies less than that of the incident beam energy as shown in Fig. 6.10. Several familiar effects can be harnessed to analyse this distribution of energies. Electric and magnetic fields, acting independently or

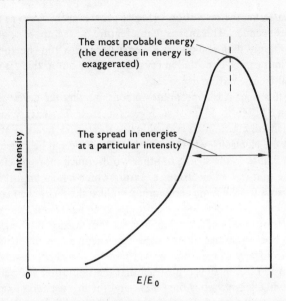

Figure 6.10 A curve illustrating the distribution of energy in an electron beam transmitted through a thin specimen. The most probable energy is reduced from the incident value and a spread of energies is introduced.

together, can be used to diperse the various electrons as a function of energy in a manner analogous to that in a mass spectrometer but here energy dispersion at constant mass rather than mass dispersion is sought. Perhaps the simplest form of energy selective system is given by the action of a sharp potential barrier created by a mesh or specially shaped electrode at the required potential. Any electron of energy less than that corresponding to the barrier will be 'reflected' by the electric field of the barrier and all energies greater than this will be transmitted as illustrated in Fig. 6.11 (a). The barrier is usually held at the same negative potential as the electron gun and biased by, say, $+V_b$ volts: any electron which has lost energy less than V_b eV is transmitted by the barrier. This system is not a dispersive one and is only crudely selective in that only a lower limit to the energy can be defined. It perhaps finds its main application in diffraction applications where removal of inelastic electrons improves the definition and the quantitative measure of the intensities.

In the case of simultaneous electric and magnetic fields arranged in a 'crossed' orientation i.e. the electric field X, magnetic field B and electron velocity v arranged in a right orthogonal orientation, it is possible by changing X or B to select electrons with a very narrow bandwidth of energy for entry into a narrow slit before photographic or electronic detection. The condition $X = v \wedge B$

ENERGY ANALYSIS AND ENERGY ANALYSING MICROSCOPES 149

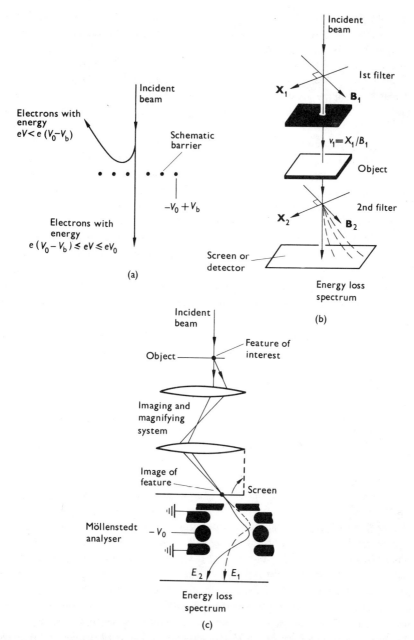

Figure 6.11 Schematic diagrams of several energy selective arrangements, (a) an electric barrier, (b) the Wien filter, (c) the Möllenstedt filter.

corresponds to zero deflection for a velocity v. This is the principle of the Wien filter illustrated schematically in Fig. 6.11 (b). A Wien filter can also be placed between the gun and specimen in an electron optical column to act as a monochromator for the incident beam. A 'double' Wien filter thus acts as a very efficient energy analysing system. In the schematic diagrams of Fig. 6.11 it can be seen that energy analysing attachments are often placed below the viewing screen of CTEMs to afford an analysis of a particular feature observed in the CTEM image.

Energy dispersion can be obtained using electric or magnetic fields with a 'prismatic' action. This is a type of analyser used in the STEM shown in Fig. 6.9. In the case of a magnetic 'prism' the orbit of the electron is given by a radius $r = 2E/F_L$ where E is the kinetic energy of the electron and $F_L = -e(v \wedge B)$ is the Lorentz force acting on the electron. The dispersion is obtained as a spread of orbit radii and any energy can be selected for detection.

A popular form of energy analyser is the Möllenstedt system shown in Fig. 6.11 (c). It consists of an arrangement of electrodes, or a magnetic equivalent, having lens properties. The action of the analyser is to accentuate the effect of chromatic aberration on the trajectories of electrons passing through the electric field of the 'lens'. The two cylindrical electrodes are at the same potential as the gun and the four plate shaped electrodes are at earth potential. The final trajectory of the electrons depends on their energy and the geometry and spacing of the electrodes in the analyser. Typical trajectories for two energies E_1 and E_2 are drawn in Fig. 6.11 (c). Electrons passing through particular areas of the specimen can be selected for energy analysis by moving the specimen so that the required area in the highly magnified image on the screen coincides with the narrow entrance slit to the analyser just below the screen. Electrons of different energy are dispersed by the analyser so that a line of electrons of varying energies is spread into a band. Typically, details of the image down to about 10 nm can be recognized in the energy spectrum and a resolution of about 2 eV at 100 kV operating voltage is obtainable.

The results of a typical problem in energy analysis are shown in Fig. 6.12. Fig. 6.12 (a) shows the position of the slit on the image of the specimen; it intercepts a well transmitting silica precipitate and part of the nickel matrix. The energy loss spectrum of the electrons passing through the slit from the silica precipitate is reproduced in Fig. 6.12 (b). The bright line on the extreme left corresponds to the electrons at 100 kV which have lost no energy and the energy loss increases with increasing distance from this line towards the right. Ideally, lines of constant energy in the spectrum should be straight, but asymmetries in the field within the analyser cause them to be curved. A densitometer trace of the photographic plate of Fig. 6.12 (b) is shown in Fig. 6.12 (c). Several maxima

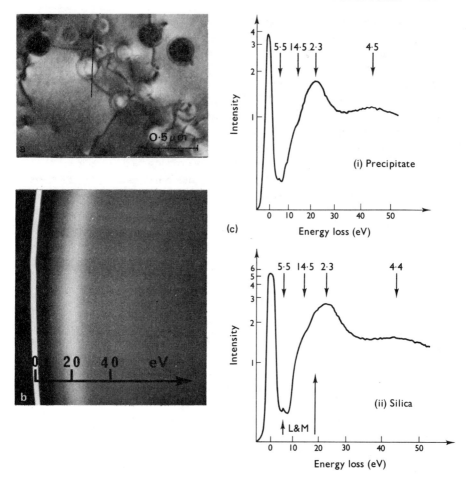

Figure 6.12 Results from an energy loss study; (a) a micrograph of the specimen with the superposed analyser slit, (b) the energy loss spectrum from the precipitate and (c) a photodensitometer trace of (b). (Courtesy of S. L. Cundy and P. J. Grundy, *Phil Mag.*, 1966, **14**, 1233.)

are evident in the loss spectrum even though the one at 45 eV is very broad and the shoulder at 14·5 eV is only just resolved. For identification the loss spectrum must be compared with some standard spectrum obtained from known material. A standard obtained from a thinned slice of silicon oxidized to silicon dioxide in steam is included in the figure; it can be inferred from the agreement between the two curves that the precipitate is composed of silicon dioxide. In many cases the loss spectrum from more than one phase is obtained at the same time i.e. one phase may be embedded in the other, and care is needed in the interpretation of the results.

In certain analysing systems, e.g. the STEM of Fig. 6.9, it is possible to form images using only 'lossy' electrons and indeed to obtain images from only a very narrow band of energies. In the CTEM the energy selector is placed between the objective and the intermediate lenses so that the magnifying system of the microscope can give the required 'lossy' micrograph.

This discussion of energy analysis is sufficient to bring out the importance of the technique. It occupies a place in the rapidly growing field of electron spectroscopy and after a development period when it has been used as a rather specialized research tool the fitting of analysing attachments to the new STEM class of microscopes is likely to become common. Texts and papers dealing with energy analysis and its applications are listed in the bibliography and Metherell (1970) has written a lengthy and authoritative review paper on the state of the art at that time.

6.6 Conclusion and future trends

Following a period of some stability the last few years have by contrast seen a tremendous upsurge in the design and development of electron optical devices. This activity is likely to continue, at least in the near future, which makes any predictions about the subject somewhat speculative. In such circumstances therefore perhaps the best policy is to review recent developments and thence look for trends.

The major stimulus to recent trends has probably been the introduction of the scanning electron microscope. In its rather unimaginative role as a 'super optical' microscope, as opposed to more specialized uses, the SEM is rapidly becoming a routine workhorse in a wide variety of applications. It is to satisfy this market that many manufacturers are producing low kilovoltage, low cost, 'table top' machines with the emphasis on ease of operation and there is every likelihood that the next few years will see these instruments taking countless emissive mode micrographs. The prospect of a 'microscopy for the million' era may not appeal to the traditionalists for whom the subject still retains esoteric virtues. It is a quirk of fate that the natural routine successor to the optical microscope did not appear until several decades of operation of the more technically demanding CTEM.

What are the features of the SEM which have caught the imagination and in so doing influenced the subsequent thinking about electron microcsopes? The most important are the wealth and variety of information derivable from a single

CONCLUSION AND FUTURE TRENDS 153

instrument and secondly, again arising from the scanning action, the opportunity for data processing. As a good example might be quoted the combination of X-ray microanalysis and emissive mode imaging coupled with storage and data handling software. In this context the straight CTEM suddenly looked rather naked and provision for conversion into an analytical microscope has since been offered by some manufacturers. This may involve the addition of crystal spectrometers or more usually the deployment of an energy dispersive detection system. Regardless of the particular system used the advantages of microanalysis have proved so effective in the study of materials, as well as of biological matter, that this refinement seems to have an assured future. Energy analysers have also been fitted to the CTEM in order to obtain information from inelastically scattered electrons which otherwise would be lost. One final development to the CTEM which should be mentioned is that of the extension beyond normal operating voltages thus allowing the examination of thicker specimens. Several laboratories are equipped with commercial 1 MV microscopes. However contrary to earlier thinking it now appears that the routine CTEM will still retain an upper limit of about 100 kV.

The disadvantage of the SEM is its comparatively poor resolution which in ordinary imaging modes nowhere approaches that of the CTEM. This is a serious limitation in certain structural studies. The attraction of marrying the resolution of the CTEM with the versatility and signal processing of the SEM is obvious and this explains the growing interest in scanning transmission microscopy. The future for the specialist STEM in the physical sciences is difficult to predict at present: its impact on biological investigation seems more certain. Some observers hold the potential of the STEM in such esteem that they predict the eclipse (whether total or partial is not clear) of conventional transmission microscopy. Supporters of this view emphasize that the STEM can duplicate the most important asset of the CTEM, namely its ability to image defects, thus making the latter obsolete. On the other hand the STEM itself may be just one stage along the path towards an all embracing multi-purpose electron microscope i.e. a device which is capable of operating in CTEM, STEM, SEM (including all the subsidiary techniques such as back scattered and luminescent image formation) and X-ray microanalysis modes. Considerable progress towards this goal has already been made by certain manufacturers.

Details of the 'packaging' of electron microscopes are really beyond the scope of this book and will be determined by the appropriate market conditions. It is clear that many of the various facilities for studying materials which have appeared recently on one machine or another will continue in demand. The benefits which have accrued from electron microscopy are becoming so well known and recognized that the future of the techniques is assured.

Appendix

We consider here information that is useful in the analysis of electron diffraction patterns obtained in the CTEM and STEM. The notes are only brief and a more complete discussion can be found in larger texts e.g. Pinsker (1953), Hirsch et al. (1965), Andrews et al. (1967).

A.1 The space lattice and its notation

Some crystal systems are more widespread in nature than others. In the particular case of metals certain higher symmetry space lattices e.g. cubic, hexagonal and tetragonal are far more common than the others.

The axes of a space lattice can be labelled x, y and z. Repetition lengths (the distance between lattice points) are usually written as a, b, and c as shown in Fig. A.1 and the relative inclinations of the axes as α, β and γ.

If a point, not necessarily a lattice point, in the unit Bravais cell is at distances from the origin of x_1, y_1, z_1, along the three reference axes, then its position is given by uvw where $u = x_1/a$ etc. As an example the coordinates of point P in Fig. A.1 are obtained as $x_1 = a/2$, $y_1 = b/2$, $z_1 = 0$. $\therefore u = \frac{1}{2}$, $v = \frac{1}{2}$, $w = 0$ and point P, in the face centre is defined by $uvw = \frac{1}{2}\frac{1}{2}0$. A lattice point can be represented by the vector $r = ua + vb + wc$ where u, v, w are the coordinates of the point. The direction of the line joining two lattice points, e.g. $u_1v_1w_1$ to $u_2v_2w_2$ is given as $[uvw]$ where $u = u_1 - u_2, v = v_1 - v_2, w = w_1 - w_2$ with u, v and w adjusted to be integers. Where negative signs occur they are placed above the integer. As an example consider the line joining the point ¼ ¼ 0 to ½ ⅓ 1; this has components $-¼, -1/12, -1$ and is written as $[\bar{3}\ \bar{1}\ \overline{12}]$ (i.e. the components are made integral by multiplying them by 12 in this case). The direction from ½ ⅓ 1 to ¼ ¼ 0 is $[3\ 1\ 12]$. Equivalent sets of directions are written as $\langle uvw \rangle$.

If a plane in a lattice intercepts the reference axes at pa, qb and rc then the plane is represented by (hkl) where

$$h : k : l = \frac{1}{p} : \frac{1}{q} : \frac{1}{r}$$

THE SPACE LATTICE AND ITS NOTATION 155

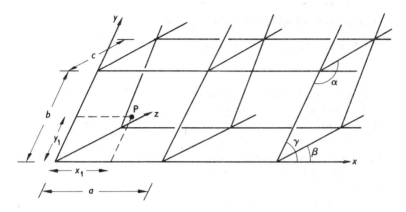

Figure A.1 A representation of a section of a space lattice showing repetition distances and interaxial angles.

Equivalent sets of planes are denoted by $\{h\ k\ l\}$; h, k, and l are called Miller indices. For example, the plane marked in Fig. A.2 has indices $3/2 : 1 : 2$ which are transformed to (3 2 4) on conversion to integers.

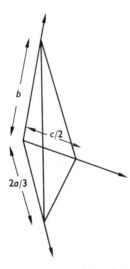

Figure A.2 A plane making intercepts $2a/3$, b and $c/2$ with the three axes of a space lattice.

A.2 Useful geometrical relationships in a space lattice

(a) The direction $[uvw]$ is in a plane (hkl) if

$$hu + kv + lw = 0$$

(b) The plane (hkl) contains two directions $[u_1 v_1 w_1]$ and $[u_2 v_2 w_2]$ if
$h : k : l = (v_1 w_2 - v_2 w_1) : (w_1 u_2 - w_2 u_1) : (u_1 v_2 - u_2 v_1)$

(c) The direction $[uvw]$ lies in both planes $(h_1 k_1 l_1)$ and $(h_2 k_2 l_2)$ if
$u : v : w = (k_1 l_2 - k_2 l_1) : (l_1 h_2 - l_2 h_1) : (h_1 k_2 - h_2 k_1)$

(d) Two planes $(h_1 k_1 l_1)$ and $(h_2 k_2 l_2)$ or directions $[u_1 v_1 w_1]$ and $[u_2 v_2 w_2]$ are at right angles if $h_1 h_2 + k_1 k_2 + l_1 l_2 = 0$ or $u_1 u_2 + v_1 v_2 + w_1 w_2 = 0$ respectively

(e) In the cubic system the direction $[hkl]$ is perpendicular to the plane (hkl)

A.3 Interplanar spacings and angles

In the interpretation of diffraction patterns it is often necessary to calculate the distance, d, between adjacent parallel planes with Miller indices (h, k, l) or (nh, nk, nl) (n is an integer).

Only the formulae for the cubic, hexagonal and tetragonal systems are given here, those for the other crystal classes are given in the references.

(a) Cubic $\dfrac{1}{d^2} = \dfrac{h^2 + k^2 + l^2}{a^2}$

(b) Tetragonal $\dfrac{1}{d^2} = \dfrac{h^2 + k^2}{a^2} + \dfrac{l^2}{c^2}$

(c) Hexagonal $\dfrac{1}{d^2} = \dfrac{4}{3} \cdot \dfrac{h^2 + hk + k^2}{a^2} + \dfrac{l^2}{c^2}$

It is sometimes useful to check the interpretation of a diffraction pattern or index a pattern by calculating the angle ϕ between crystal planes. This is not usually necessary but the formula for the cubic system is

$$\cos \phi = \frac{h_1 h_2 + k_1 k_2 + l_1 l_2}{[(h_1^2 + k_1^2 + l_1^2)(h_2^2 + k_2^2 + l_2^2)]^{1/2}}$$

Equations for other crystal symmetries are given in the references.

A.4 Details of the reciprocal lattice

Of interest here is the relevance of the reciprocal lattice to the solution of electron diffraction patterns and in the explanation of electron diffraction

STRUCTURE FACTOR 157

effects and contrast. The following points are noteworthy: —

(a) In a diffraction pattern it is found that distances are inversely proportional to distances in the real crystal lattice (see Section 2.2.2). It is therefore convenient to introduce a lattice which is related in an inverse sense to the real lattice; this is the reciprocal lattice defined by vectors a^*, b^* and c^* such that $a^* . a = b^* . b = c^* . c = 1$ and $a^* . b = b^* . c = c^* . a = a^* . c = 0$ where a, b and c are the primitive translation vectors of the real lattice (see A.1).

(b) For orthogonal axes $|a| = 1/|a^*|, |b| = 1/|b^*|, |c| = 1/|c^*|$ and a^* is parallel to a; b^* to b and c^* to c.

(c) The reciprocal lattice vector

$$g_{hkl} = ha^* + kb^* + lc^*$$

is perpendicular to the real crystal plane with Miller indices hkl and $|g_{hkl}| = 1/d_{hkl}$ where d_{hkl} is the spacing of the real crystal planes. A reciprocal lattice point hkl lies in the (uvw) reciprocal lattice plane if $hu + kv + lw = 0$ (this applies to all crystal classes).

A.5 Structure factor

In X-ray and electron diffraction some diffraction rings or spots are forbidden because of the atomic arrangement within the unit Bravais cell (see e.g. Brown and Forsyth (1973)).

The intensity I scattered from a unit cell is summed over all the atom positions in the unit cell to take account of phase relationships. This intensity is proportional to F^2, where F is the 'structure factor' of the unit cell given by

$$F = \Sigma_i f_i \, e^{-2\pi i g . r_i}$$

f_i is the amplitude scattered from atom $i, g = ha^* + kb^* + lc^*, r_i = u_i a + v_i b + w_i c$ and $e^{-2\pi i g . r_i}$ is a phase term. Remembering $a^* . a = 1$ and $I \propto FF^*$ then

$$I \propto [\Sigma_i f_i \cos 2\pi(hu_i + kv_i + lw_i)]^2 + [\Sigma_i f_i \sin 2\pi(hu_i + kv_i + lw_i)]^2$$

The second term of the above equation is always zero in a centro-symmetric crystal because $\sin n\pi = 0$. For planes (hkl) for which $I = 0$ no diffraction can occur and Bragg reflections are absent.

Example In a body centred cubic unit cell the atomic positions can be given coordinates $u, v, w, = 0,0, 0$ or ½, ½, ½ then $I = f^2 [1 + \cos \pi(h + k + l)]^2$ and $h + k + l$ must always be an even integer for $I \neq 0$. Hence planes with $(hkl) = (100), (111), (210)$ etc. have no associated Bragg reflection.

A.6 Indexing a polycrystalline ring pattern

In a diffraction pattern from a polycrystalline specimen all possible reflections (hkl) save those forbidden by the structure factor are present because all the planes (hkl) are to be found somewhere in the specimen at the Bragg angle to the incident beam. Rings are formed because throughout the specimen any one type of plane will take all azimuth angles to the incident beam. As the grain or crystallite size increases and approaches the size of the beam, the rings break down into spots until the grain size is greater than the beam spot when a single crystal pattern is obtained.

A pattern from a specimen showing a fibre 'texture' has certain reflections missing because certain axes or planes are parallel or perpendicular to the beam direction respectively.

Suggested procedures for the indexing of diffraction patterns are as follows. If the camera constant λL is known then measurement of the D values (see Fig. A.3) from the pattern will yield the d spacings through the relation $d_{hkl} = \lambda L/D_{hkl}$. Provided the crystal structure and lattice constants are known, equations such as those contained in Section A.3 can now be used to determine (hkl). A similar procedure is possible when λL is not known; this time however the ratios of D values are used (i.e. for two rings $D_1/D_2 = d_2/d_1$) and the probable values of d obtained by calculation can be compared with this ratio and

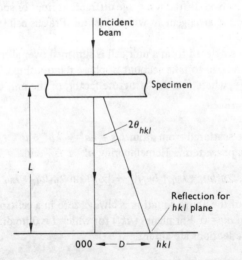

Figure A.3 Illustrating the relationship between the effective camera length, the Bragg angle and the distance between the reflection and the centre of the diffraction pattern.

if necessary confirmed by another ratio $D_1/D_3 = d_3/d_1$. In a more general case where the crystal structure is not known computer procedures are best used. However, once λL has been found the d spacings of any material can be found and a check against the ASTM (see References) files may produce a positive result.

For certain Bravais lattices the characteristic absent reflections imposed by the structure factor may provide a strong clue to indexing. Thus a b.c.c. pattern consists of rings whose radii are proportional to 2, 4, 6, 8 etc. In an f.c.c. crystal h, k and l are all odd or all even.

A.7 Plotting a single crystal pattern

It is often convenient to plot out the simplest single crystal spot patterns for direct comparison with observed patterns. The simplest patterns are obtained when the incident beam is normal to prominent crystallographic planes e.g. $\{100\}$ or $\{111\}$ in the lattice. If the incident beam is parallel to the $[uvw]$ direction in the crystal the diffraction pattern produced will correspond to a plane of the reciprocal lattice perpendicular to the incident beam (Section 2.2.2). From Section A.4 it will be seen that the appropriate reciprocal lattice plane is the (uvw) plane i.e. the (uvw) reciprocal plane is always perpendicular to the real crystal $[uvw]$ direction; this plane may be deduced as follows.

Using the first result in Section A.2 choose any two spots $h_1 k_1 l_1$ and $h_2 k_2 l_2$ to satisfy

$$h_1 u + k_1 v + l_1 w = 0$$
$$h_2 u + k_2 v + l_2 w = 0$$

It follows that $h_1 + h_2$, $k_1 + k_2$ and $l_1 + l_2$ also satisfy this condition. Calculate d_{hkl} and D_{hkl} for these spots, Section A.6, (if λL is not known assume a constant convenient value, e.g. 5 nm mm). Draw a triangle with the calculated sides. Complete the regular cross-grating patterns by addition of vectors as in Fig. A.4.

Example To plot the $(uvw) = (310)$ reciprocal lattice plane for an f.c.c. crystal.

By inspection $(h_1 k_1 l_1) = (002)$ is present in this plane because $h_1 u + k_1 v + l_1 w = 0$. A perpendicular spot $h_2 k_2 l_2$ must satisfy the condition $h_1 h_2 + k_1 k_2 + l_1 l_2 = 0$ and also $h_2 u + k_2 v + l_2 w = 0$, i.e. $l_2 = 0$ and $3h_2 + k_2 = 0$, or $h_2 k_2 l_2 = (1\bar{3}0)$. This reflection is forbidden in the f.c.c. system (hkl must be all odd or all even, 0 counting as even) but $(2\bar{6}0)$ is allowed. Draw the basic triangle as described above and shown in Fig. A.5. Draw in $(2\bar{6}2)$ and also $(1\bar{3}1)$ at half the $D_{2\bar{6}2}$ value. Fill in the other spots by symmetry.

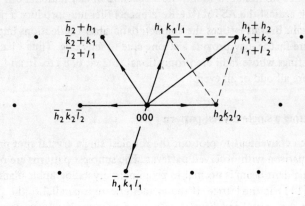

Figure A.4 A plot of part of a plane of the reciprocal lattice or a single crystal spot pattern.

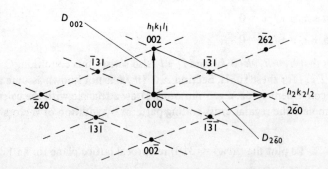

Figure A.5 A construction of the (310) reciprocal lattice plane or [310] diffraction pattern for an f.c.c. crystal.

Indexing a single crystal spot pattern

The initial indexing procedure is similar to that described for polycrystalline patterns and general indices (hkl) are tentatively assigned to each spot. However several permutations of (hkl) may give rise to identical D values, e.g. in the cubic system all spots with the same $h^2 + k^2 + l^2$ have a common D value. Therefore choose three low order spots (i.e. near to the centre) as shown in Fig. A.4. From above we know that $h_3 = h_1 + h_2$, $k_3 = k_1 + k_2$ and $l_3 = l_1 + l_2$ and any ambiguity can be removed. A check on the indexing is supplied by measuring the angle on the pattern between two arbitrary reciprocal lattice vectors, say $g_1 = h_1 a^* + k_1 b^* + l_1 c^*$ and $g_2 = h_2 a^* + k_2 b^* + l_2 c^*$. This angle is also that between the $(h_1 k_1 l_1)$ and $(h_2 k_2 l_2)$ planes in the crystal and can be calculated (Section A.3) or perhaps found in tables. If the measured and calculated angle values agree, the indexing may be presumed correct.

To find the direction $[uvw]$ in the crystal to which the incident beam is parallel we can use the results that $h_1 u + k_1 v + l_1 w = h_2 u + k_2 v + l_2 w = 0$ whence (Section A.2 (c))

$$u = k_1 l_2 - l_1 k_2 ; v = l_1 h_2 - h_1 l_2 ; w = h_1 k_2 - k_1 h_2$$

A.8 Kikuchi lines and their use in the determination of orientation

(a) Exact reflecting position

A crystal is in the exact reflecting position for a reflection g (i.e. (hkl) is at exactly the Bragg angle) when the reflecting sphere passes through the spot. In this case the Kikuchi lines pass through 000 and hkl respectively as in Fig. A.6 (a).

(b) Deviation from the reflecting position

When the zone axis or $[uvw]$ is exactly parallel to the beam the Kikuchi lines are halfway between O and X, Fig. A.6 (b). The deviation $\Delta\theta$ from the reflecting position is normally expressed as a distance proportional to $\Delta\theta$ i.e.

$$\frac{\Delta X}{OX} = \frac{\Delta\theta}{2\theta}$$

(c) Determination of specimen tilt from moving Kikuchi lines

A diffraction pattern is photographed before and after the crystal is tilted in the microscope. Any Kikuchi lines present move across the pattern, i.e. AB in Fig. A.6 (c) moves to A' B' and CD moves to C' D'.

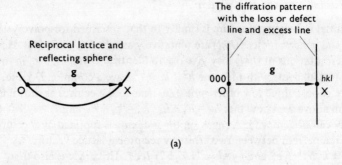
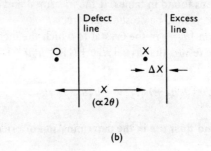
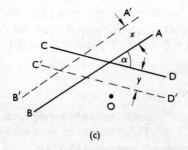

Figure A.6 The relationship between the position of Kikuchi lines and diffraction (a) at the exact Bragg position, (b) at a deviation $\Delta\theta$ and (c) after tilting the crystal.

If the distances moved are x and y respectively then the angle of tilt of the crystal is

$$\Delta\theta = \frac{PQ}{L}$$

(L is the camera length) and

$$PQ = \frac{(x^2 + y^2 + 2xy \cos \alpha)^{1/2}}{\sin \alpha}.$$

References

Amelinckx, S., Gevers, R., Remaut, G. and Van Landuyt, J. (eds) 1970. *Modern Diffraction and Imaging Techniques in Material Science*, North-Holland, Amsterdam and London.
Andrews, K. W., Dyson, D. J. and Keown, S. R. 1968. *Interpretation of Electron Diffraction Patterns*, Adam Hilger, London.
Brammar, I. S. and Dewey, M. A. P. 1966. *Specimen Preparation for Electron Metallography*, Blackwell, Oxford.
Cochran, W. 1973. *The Dynamics of Atoms in Crystals*, Edward Arnold, London.
Coles, B. R. and Caplin, A. D. 1976. *The Electronic Structures of Solids*, Edward Arnold, London.
Crewe, A. V. 1970. *Quarterly Reviews in Biophysics*, **3**, 137.
Ditchfield, R. W., Grubb, D. T. and Whelan, M. J. 1973. *Phil. Mag.* **27**, 1267.
Doyle, P. A. and Turner, P. S. 1968. *Acta Cryst.* **A24**, 390.
Dugdale, J. S. 1976. *The Electrical Properties of Metals and Alloys*, Edward Arnold, London.
Everhart, T. E. and Thornley, R. F. M. 1960. *J. Sci. Inst.* **37**, 246.
Fathers, D. J., Joy, D. C. and Jakubovics, J. P. 1973. *Scanning Electron Microscopy: Systems and Applications* 1973, 214, Institute of Physics, London and Bristol.
Ferrier, R. P. 1969. *Advances in Optical and Electron Microscopy*, **3**, 155.
Glauert, A. M. 1972. *Practical Methods in Electron Microscopy*, North-Holland, Amsterdam and London.
Goringe, M. J. 1973. *J. Microscopy*, **97**, 95.
Grundy, P. J. and Tebble, R. S. 1968. *Advances in Physics*, **17**, 153.
Haine, M. E. and Cosslett, V. E. 1961. *The Electron Microscope*, Spon, London.
Hawkes, P. W. 1972. *Electron Optics and Electron Microscopy*, Taylor and Francis, London.
Hearle, J. W. S., Sparrow, J. T. and Cross, P. M. (ed) 1972. *The Use of the Scanning Electron Microscope*, Pergamon Press, Oxford and London.
Heidenreich, R. D. 1964. *Fundamentals of Transmission Electron Microscopy*, Interscience, Wiley, New York.
Henderson, B. 1972. *Defects in Crystalline Solids*, Edward Arnold, London.
Hirsch, P. B., Howie, A., Nicholson, R. B., Pashley, D. W. and Whelan, M. J. 1965. *Electron Microscopy of Thin Crystals*, Butterworths, London.

REFERENCES

Jacobs, M. H. 1974. *Advances in Analysis of Microstructural Features by Electron Beam Techniques*, 80, The Metals Society.
Kay, D. H. (ed) 1965. *Techniques for Electron Microscopy*, Blackwell, Oxford.
Lorimer, G. W., Razik, N. A. and Cliff, G. 1973. *J. Microscopy*, **99**, 153.
Makin, M. J. and Sharpe, J. V. 1968. *J. Mater. Sci.* **3**, 360.
Metherall, A. J. F. 1970. *Advances in Optical and Electron Microscopy*, **4**, 263.
Oatley, C. W. 1972. *The Scanning Electron Microscope – Part 1. The Instrument*, Cambridge University Press.
Sandstrom, R., Spencer, J. S. and Humphreys, C. J. 1974. *J. Phys. D.* **7**, 1030.
Shaw, D. A. and Thornton, P. R. 1968. *J. Mater. Sci.* **3**, 507.
Thornton, P. R. 1968. *Scanning Electron Microscopy*, Chapman and Hall, London.
Valdre, U. (ed) 1971. *Electron Microscopy in Material Science*, Academic Press, London.
Wells, O. C. 1974. *Scanning Electron Microscopy*, McGraw-Hill, New York.
Wells, O. C., Broers, A. N. and Bremer, C. G. 1973. *Appl. Phys. Lett.* **23**, 353.
Yeoman Walker, D. E. 1973. *Scanning Electron Microscopy: Systems and Application 1973*, 202, Instiue of Physics, London and Bristol.

Bibliography

Below are categorized a number of books which may be useful for further reading. Some have already been listed in the references but are included for the sake of completeness.

1. *History of Microscopy*
   Ford, B. J. 1973. *The Revealing Lens*, Harrap, London.
   Marton, L. 1968. *Early History of the Electron Microscope*, San Francisco Press Inc., San Francisco.

2. *Electron Diffraction and Scattering Theory*
   Andrews, K. W., Dyson, D. J. and Keown, S. R. 1968. *Interpretation of Electron Diffraction Patterns*, Adam Hilger, London.
   Ball, C. J. 1971. *Introduction to the Theory of Diffraction*, Pergamon Press, Oxford.
   Guinier, A. 1963. *X-ray Diffraction in Crystals, Imperfect Crystals and Amorphous Bodies*, Freeman, San Francisco and London.
   Pinsker, Z. G. 1953. *Electron Diffraction*, Butterworths, London.
   Rymer, T. B. 1970. *Electron Diffraction*, Methuen, London.
   Vainshtein, B. K. 1964. *Structure Analysis by Electron Diffraction*, Pergamon Press, Oxford.

3. *Electron Optics and Electron Microscopes*
   Grivet, P. 1972. *Electron Optics*, Pergamon Press, Oxford.
   Haine, M. E. and Cosslett, V. E. 1961. *The Electron Microscope*, Spon, London.
   Hawkes, P. W. 1972. *Electron Optics and Electron Microscopy*, Taylor and Francis, London.
   Klemperer, O. and Barnett, M. E. 1971. *Electron Optics*, Cambridge University Press.
   Oatley, C. W. 1972. *The Scanning Electron Microscope – Part 1. The Instrument*, Cambridge University Press.
   Paszkowski, B. 1968. *Electron Optics*, Iliffe, London.

Swift, J. A. 1970. *Electron Microscopes*, Kogan Page, London. Barnes and Noble, New York.
In addition there are specialized journals which deal with this topic e.g. *Advances in Optical and Electron Microsocpy*, (Academic Press).

4. *Applications of Electron Microscopy*

Amelinckx, S., Gevers, R., Remaut, G. and Van Landuyt, J. (eds) 1970. *Modern Diffraction and Imaging Techniques in Material Science*, North-Holland, Amsterdam and London.

Belk, J. A. and Davies, A. L. (eds) 1968. *Electron Microscopy and Microanalysis of Metals*, Elsevier, Amsterdam, London and New York.

Edington, J. W. 1974/75. *Practical Electron Microscopy in Materials Science*, Macmillan, London.

Glauert, A. M. (ed). 1972, etc. *Practical Methods in Electron Microscopy*, North-Holland, Amsterdam and London.

Hall, C. E. 1966. *Introduction to Electron Microscopy*, McGraw-Hill, New York and London.

Hearle, J. W. S., Sparrow, J. T. and Cross, P. M. (eds) 1972. *The Use of the Scanning Electron Microscope*, Pergamon Press, Oxford.

Heidenreich, R. D. 1964. *Fundamentals of Transmission Electron Microscopy*, Interscience, Wiley, New York.

Hirsch, P. B., Howie, A., Nicholson, R. B., Pashley, D. W. and Whelan, M. J. 1965. *Electron Microscopy of Thin Crystals*, Butterworths, London.

Holt, D. B., Muir, M. D., Grant, P. R. and Boswarva, I. M. (eds) 1974. *Quantitative Scanning Electron Microscopy*, Academic Press, London.

Kay, D. H. (ed) 1965. *Techniques for Electron Microscopy*, Blackwell, Oxford.

Swann, P. R., Humphreys, C. J. and Goringe, M. J. (eds) 1974. *High Voltage Electron Microscopy*, Academic Press, London.

Thomas, G. 1964. *Transmission Electron Microscopy of Metals*, Wiley, New York and London.

Thornton, P. R. 1968. *Scanning Electron Microscopy*, Chapman and Hall, London.

Valdré, U. (ed) 1971. *Electron Microscopy in Materials Science*, Academic Press, London.

Wells, O. C. 1974. *Scanning Electron Microscopy*, McGraw-Hill, New York.

There are also many conferences on electron microscopy whose proceedings are published. In particular might be mentioned the *Proceedings of the Annual Scanning Microscopy Symposium* published by IITRI, Chicago (ed. Johari) which deals with SEM and STEM, and the *Proceedings of the International and European Congresses in Electron Microscopy*. Several journals contain research papers concerned with the applications of electron microscopy e.g. *Philosophical Magazine, Journal of Materials Science, Physica Status Solidi* and *Journal of Microscopy*.

Index

α-fringes, 91
Abbé, 1, 26
aberrations, 42–5
  astigmatism, 45, 57
  chromatic, 44, 57, 132, 141
  spherical, 43, 56, 93, 132, 141
  and resolution, 56–61, 132–3
absorptive mode, 99, 113, 125
absorbed currents, 38, 113
Airy disc, 1, 44
amplitude,
  contrast (see contrast)
  contrast transfer function, 93
  phase diagram, 80
  scattered from crystal, 68–78
  specimen, 2
analytical TEM, 137–9, 152
anode, 39, 144
anomalous absorption, 72
antiphase boundaries, 91
aperture,
  angle, 141
  function, 93
astigmatism, 45, 47, 57
atomic,
  displacement (see displacement damage)
  number contrast, 32, 102
  potential, 12
  scattering (see scattering),
Auger,
  electrons, 36, 115
  mode, 99, 125
  spectroscopy, 113
avalanche breakdown, 124

backscattered electrons (see reflected electrons)
band structure, 35, 121
beam,
  induced conductivity, 38, 118–24
  chopping coils, 53, 126
  coherence, 4, 47
  divergence, 46, 52
  spreading, 29, 58, 137
bend extinction contour, 77
Bloch waves, 68, 108
Born approximation, 13, 72
Bragg,
  law, 19
  angle, 19
  reflection, 19, 47, 107, 142
Bravais lattice, 154
bright field, 49–50, 142–3
brightness (see electron gun)
Brillouin zone boundary, 69
bubble domain, 97
Burgers vector, 82–4

calibration,
  camera length, 21, 51
  image rotation, 45
  magnification, 47
camera, 47, 53
  constant, 158, 159
  length, 19, 21, 51
carrier lifetime (see lifetimes)
cathode, 39, 41
cathodoluminescence, 35, 115–8, 125
charge collection (see beam induced conductivity)
chromatic aberration, 44, 57, 132, 141
Cockcroft–Walton generator, 132
coherency, 86
coherence length, 4, 47
collective losses, 26
collector for SEM, 105, 107
  Everhard–Thornley, 54–5
  solid state, 55, 125
column approximation, 80, 129
conductive mode, 99, 118–24
contamination, 39
contrast in CTEM, 16–26, 62–97, 128–39
  at defects, 78–91, 134–7
  deficiency, 16
  diffraction, 62–97, 128–39
  phase, 25–6, 92–7

172   INDEX

contrast in SEM, 34–8, 59, 99
  absorbed current, 113
  atomic number, 32, 102
  conductive, 118–24
  electric & magnetic field, 35, 105
  luminescent, 115–8
  topographic, 34, 99–107
  voltage, 105–7
contrast in STEM, 142
critical mass thickness, 16
crossover, 41, 45
crystal spectrometer, 113–5, 137
CTEM, 4
  design, 45 ff, 131–3 (HVEM)
  applications, 62 ff
current density, 41

dark field images,
  CTEM, 49, 91
  STEM, 142, 143
Debye equation, 14
de Broglie equation, 3
deficiency contrast, 16 ff, 68
defocussing, 93–5
depletion region, 38, 121, 122
depth of field/focus, 45, 56, 61
differential signals, 35, 126
diffraction, 17 ff, 68 ff
  aperture, 49
  contrast, 68 ff
  dynamical theory of, 68–72
  Fraunhofer, 19, 93
  kinematical theory of, 72–8
  low angle, 51
  and resolution, 1, 57, 58
  in STEM, 142
diffraction pattern, 14, 142
  indexing, 158
  plotting, 159
  ring, 158
  selected area, 51
  spot, 49, 159
diffusion length, 117, 121
disc of least confusion, 44, 58
dislocation, 83–5, 141
dislocation contrast,
  CTEM, 82–4,
  SEM, 118, 120
  STEM, 141
dispersion surface, 69
displacement,
  damage, 27, 130, 135
  energy, 131
  vector, 79 ff

display CRT, (see video CRT)
double condenser, 45–7, 65

Einstein's equation, 124
electric field contrast, 99, 105
electron diffraction (see diffraction)
electron,
  energy, 3, 8
  wavelength, 3
electron channelling patterns (ECP's),
    107–13, 125
  indexing, 110, 111
  resolution, 109
  selected area, 111
electron gun, 39–41
  brightness, 41, 144
  field emission, 41, 144–5
  thermionic emission, 40
  triode, 40
electron–hole pairs, 36, 38, 118 ff
EMMA-4, 137–9
emissive mode, 99–107, 116, 125
energy,
  analysis, 126, 144, 147–51
  dispersive X-ray analysis, 113–5
  losses, 26
environmental cell, 128, 134
Ewald (reflecting) sphere, 23, 129
extinction distance, 72, 76, 89

Fermi surface, 69
fibre texture, 158
field emission source, 144–5
filament, 39–41
focal length, 42–4, 144
Fourier transform, 93
fractography, 101
Fraunhofer diffraction, 19, 93
frequency spectrum, 93
Fresnel diffraction, 95

Gaussian,
  image plane, 43
  spot size, 51 ff, 61
g·b or invisibility criterion, 81–4, 88, 141

heating, of specimen, 27
Helmholtz, 139
high voltage TEM, 128–37
  aberrations and resolution, 131–2
  applications, 133–7
high voltage STEM, 146
hybrid microscopes, 67, 142

INDEX

image formation,
    CTEM, 10–11, 47–51
    SEM, 5, 53
    STEM, 139
image processing, 126
incoherent scattering (see scattering)
indexing of diffraction patterns, 158–63
inelastic scattering (see scattering)
intensity of diffracted beams, 68–82
    from a thin crystal, 75
    oscillations, 71, 76
interaction volume, 29, 116
interplanar angles, 156
    spacings, 156
ionization damage, 130
invisibility criterion, 81–4, 88, 141

Kikuchi lines, 21, 143
    in diffraction analysis, 161
kinematical theory of diffraction, 72–8
Kossel pattern lines, 125

LaB_6 filament, 41
Langmuir formula, 41, 60
lattice,
    defects, contrast from, 79–86, 134
    interference fringes, 26, 95
    reciprocal, 156
    vector, 154
    vibrations, 27
Laue equations, 24
lens,
    condenser, 45
    condenser–objective, 138
    magnetic, 42 ff
    mini, 137
    objective, 42, 47, 52
    projector, 42
lifetimes, of carriers, 36, 38, 117, 123, 418
light pipe, 54
line of no contrast, 88
line scan, 59, 114, 121–2
loss,
    micrographs, 151
    spectrum, 28, 150
Lorentz,
    force, 95, 105
    microscopy, 95–7
luminescent mode, 115–8, 125

magnetic,
    analyser, 148
    contrast in CTEM, 95
    contrast in SEM, 105, 126

    domains, 95, 105, 106, 126
    lens (see lens)
magnification, 55–6, 104
    calibration, 47
    raster size, 55–6, 61
mass,
    thickness contrast, 12–7
    thickness, critical, 16
    range, 30
    relativistic, 3
matrix,
    contrast, 87
    strain field, 87, 91
Miller indices, 19, 155
mini-lens, 137
modes in SEM, 99, 125
Mollenstedt analyser, 148
Moiré fringes, 90
multiple beam effects, 129

neutrality current, 38
neutron irradiation, 85, 134
noise and resolution, 59–61, 116

objective,
    aperture, 49 ff
    lens (see lens)
orientation contrast, 90

particle distribution, 62 ff
path difference, 22
periodic potential, 69
phase,
    contrast, 25
    difference, 24
    factor, 82
    image, 94–7
    object, 2
    transfer function, 93
phonon excitation, 27
plasmon,
    frequency, 26
    scattering, 26
p–n junction, 36, 107, 119 ff
p–n–p transistor, 114, 124
picture point, 59, 109
polymers, 103
polytypic bands, 116–7
precipitates, 96–90
    coherent, 86
    contrast at, 86–90
primary electrons (see reflected electrons)
probe size, 52, 55, 60, 137
projector lens (see lens)

INDEX

radiation,
  damage, 27, 110, 130
  hazard, 132
range,
  primary electron, 29–30
  extrapolated, 30
recombination processes, 35–6, 117
reciprocal lattice, 19, 156
  point, 75, 77
  vector, 22, 157
reciprocity theorem, 139–41
reflected electrons, 31–2, 35
  angular distribution, 31–2
  channelling patterns, 107–13
  dependence on Z, 32, 101–2
  energy distribution, 31
  yield, 31
refractive index, 72
relativistic accelerating voltage, 3
replicas, 65–8
residual contrast, 82
resolution (resolving power),
  definition of, 1
  of ECP.s, 109
  in CTEM, 2, 4, 56–8
  in HVEM, 128, 132–3
  in SEM, 6, 58–61, 99, 107
  in STEM, 141–6
  lattice, 57
  point to point, 57
ring pattern, 21, 158
rocking beam, 111–2, 142–3
Rutherford scattering, 13

sample volume, 29, 113
scan coils, 52, 111
scanning electron diffraction, 143
scattering,
  cross sections, 12
  elastic, 8, 10–26
  factor, 13
  from bulk material, 28–38
  high energy, 128–31
  inelastic, 8, 26–8, 133
  parameter, 13
  plasmon, 26
  Rutherford, 13
Schrödinger equation, 69
secondary electrons, 31–8
  angular distribution, 35
  collection, 54
  energy, 31
  range, 30
  yield, 33

selected area,
  diffraction, 51
  channelling pattern, 111–2
SEM
  design, 8, 51–6
  applications, 98–126
shadow,
  casting, 66
  micrographs, 63
signal processing, 126, 152
signal-to-noise ratio, 59 ff
space lattice, 154
specimen holder, 47
spherical aberration, 41, 43, 56, 58
spot pattern, 49, 159 ff
stacking fault, 81
STEM, 139 ff
  construction, 142 ff
  high resolution, 144–6
stereo pairs, 104
strain contrast, 87
structure factor, 74, 157
  contrast, 88

textiles, 103, 104
thickness fringes, 72, 77
'top–bottom' effect, 146
topographic contrast, 33–5, 99–105
transmission of electrons, 10–26, 128–31
transmissive mode (see STEM)
transparency, 129
transparency or specimen function, 93
two-beam approximation, 69

video CRT, 53, 109, 142
viewing screen, 47, 51
voltage contrast, 99, 105–7, 123

waves (1) and (2), 69
wavelength dispersive analysis, 114
weak beam technique, 91
Wehnelt cylinder, 39
Wien filter, 148
working distance, 52, 55, 104

X-ray,
  absorption, 114, 138
  characteristic spectra, 37, 137
  ED analysis, 113, 114
  fluorescence, 114, 138
  mode, 99, 125
  WD analysis, 113, 114

Y-modulation, 54